To Eve with ~~~~~

Dad.

Christmas 1985.

D1796308

Tropical and Subtropical Plants

How to recognize them

Frances Perry and Roy Hay

Ward Lock Limited · London

Acknowledgements

The publishers gratefully acknowledge the following persons, agencies and company for granting permission to reproduce the colour illustrations: the Florapic Library (all photographs except those cited below); Gillian Beckett, pp. 9 (top), 17 (top left), 19 (lower right), 71 (top left) and 101 (top left); George Hurn, p. 107 (lower); Judy Todd, pp. 91 (lower right) and 113 (top); the Harry Smith Horticultural Photographic Collection, p. 101 (top right); and Messrs. Thompson & Morgan Ltd., pp. 17 (top right) and 113 (lower left). The publishers regret that, despite considerable effort, they have been unable to identify the copyright holder of the photograph of *Brassaia actinophylla*, p. 17 (lower), to whom acknowledgement is made.

The line drawings in the pictorial glossary and the symbols were drawn by Rosemary Wise.

Cover photographs

FRONT: upper left: *Ipomoea acuminata*
upper right: *Nymphaea ampla*
lower left: *Psidium guajava*
lower right: *Plumeria rubra*
BACK: *Yucca aloifolia* 'Marginata'
All photographs courtesy Florapic except *Psidium guajava*, which is courtesy Thompson and Morgan Ltd.

© Frances Perry and Roy Hay 1982
First published in Great Britain in 1982
by Ward Lock Limited, 82 Gower Street,
London WC1E 6EQ, a Pentos Company.
All Rights Reserved. No part of this publication
may be reproduced, stored in a retrieval system,
or transmitted, in any form or by any means,
electronic, mechanical, photocopying, recording,
or otherwise, without the prior permission of the
Copyright owners.
House editor Denis Ingram
Designed by Charlotte Westbrook
Text filmset in 10 point Apollo
Colour origination by Bridge Graphics Limited, Hull
Printed and bound in Hong Kong by
South China Printing Company

British Library Cataloguing in Publication Data

Perry, Frances
 Tropical and subtropical plants
 1. Plants
 I. Title   II. Hay, Roy
 581   QK45.2

 ISBN 0–7063–6137–4
       0–7063–5964–X–Pbk

Contents

Preface

Thanks to easier forms of transport thousands of people now travel as a matter of course to different countries and other continents. Apart from many much advertised attractions, they find strange new plants and flowers, especially in the tropics and subtropics which are particularly rich in flowering trees and shrubs. Most of these cannot be grown—except under glass—in cool climates, some are even too big for that and so are never seen except by travellers. But, finding out their names is a real headache. In our experience most people when asked either profess ignorance or else give a native name difficult to track down.

Accordingly we embarked on this book in the hope that it might help others to identify some of the plants they are likely to come across in the tropics and subtropics. Every plant described we have met with in at least two continents, proving that enlightened gardeners, botanists and parks authorities recognize their worth.

Many portrayed in the following pages will be found in parks and gardens or growing as street trees in warm climates. For easy recognition we have grouped them under various headings, for example trees, shrubs, climbers, and those found in or near water.

However, one cannot be dogmatic about plants and some of the descriptions or groupings are to some extent arbitrary. A shrub, for instance, is normally a woody plant with many basal stems, but occasionally it develops a single trunk and becomes a small tree. So, if in doubt, study both groups. Heights and seasons may also vary according to the type of soil, availability of water and various climatic factors, but those we give should serve as a rough guide. The plants are listed in each group by alphabetical order of Latin names, in most cases genus and species. Occasionally the natural variety (var.) or the cultivated variety (denoted by quotation marks) are also cited.

Various people have helped in the compilation of this book and we are especially grateful to our son Roger Perry who has spent many years in tropical and subtropical countries and whose assistance has been invaluable; the staff of the Cambridge Botanic Garden and K. Beckett of King's Lynn for help with some identification queries. We are also indebted to our daughter-in-law Shirley Perry who typed all the manuscript and to various people for illustrations, who are acknowledged separately.

F.P.
R.H.

4

Key to Symbols

Ψ   Deciduous plants.

Ψ   Evergreen plants.

LEAVES

⌀   Simple leaf. Simple leaves can be various shapes, with or without toothed edges.

🜂   Entire leaf with three leaflets. Entire: with an unbroken margin, not toothed at the edge.

❋   Entire leaves growing in whorls. Whorl: a ring of leaves (or flowers) around a stem at the same level as each other.

🜊   Entire leaf, lobed. Lobed: leaves (or petals) which are divided by indentations, but not into separate leaflets.

❋   Compound, palmate leaf. Compound: divided into separate leaflets. Palmate: lobed or divided into more than three leaflets arising from a central point.

↕   Compound, pinnate leaf. Pinnate: with leaflets arranged on opposite sides of a common stalk.

⚕   Compound, bipinnate leaf. Bipinnate: where the divisions of a pinnate leaf are themselves pinnate.

FLOWERS

🞧   Simple, round flower.

❀   Daisy-like flower.

🜨   Pea-shaped flower.

✻   Inflorescences loosely arranged, often branched or in sprays, erect or drooping, flowers often small, axillary or terminal. Inflorescence: the arrangement of the flowers on a stem or branch. Axil: the upper angle formed at junction of stem and leaf. Terminal: borne at the top of the stem.

◁   Funnel-shaped flower, sometimes flaring at mouth.

⊐   Tubular flower, long and narrow.

◖   Spathe, arum-like flower. Spathe: a large bract enclosing a flower cluster, the spadix.

↰   Irregular flower, one that is not symmetrical, may be lopsided or lipped.

◇   Flowers tightly packed in round heads, cones or umbels.

◠   Bell-shaped flower, erect or pendent.

occ. occasionally.
m   mostly.

5

Pictorial Glossary

LEAVES

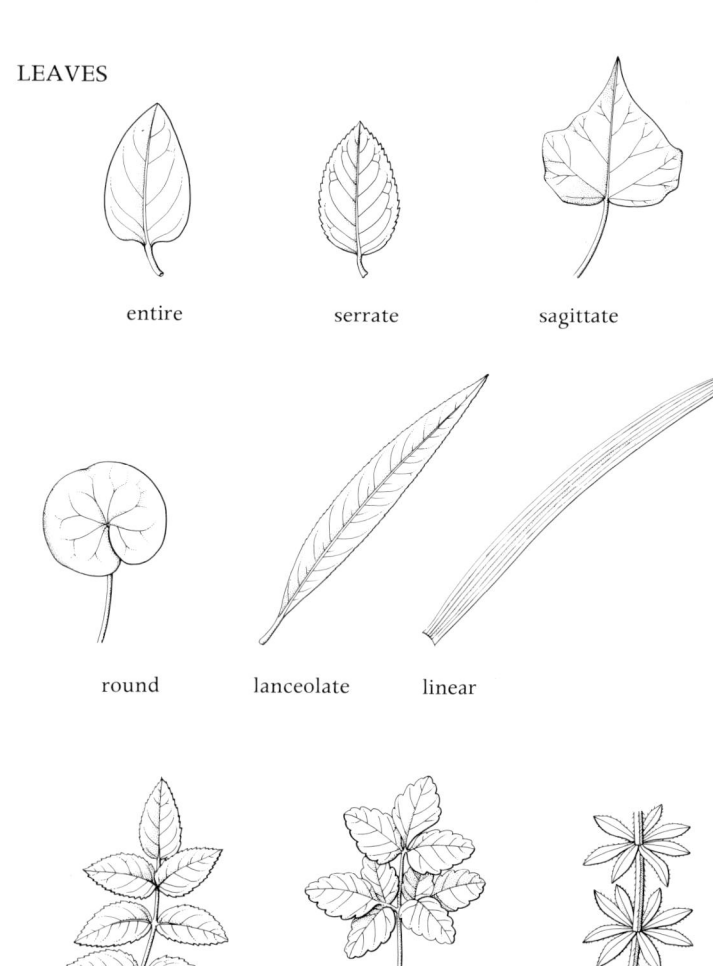

entire

serrate

sagittate

round

lanceolate

linear

pinnate

bipinnatc

whorls

palmate

opposite leaves

alternate leaves

FLOWERS

solitary

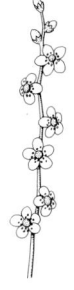

spike

raceme

panicle

umbel

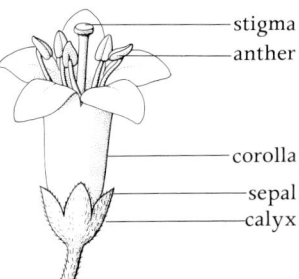

— stigma
— anther
— corolla
— sepal
— calyx

structure of tubular flower

7

1
Trees

Acacia mearnsii   Ψ ⚘ ❄
Black wattle; mimosa

Family *Leguminosae*   **Subfamily** *Mimosoideae*
Place of origin Tasmania, Australia
Description There are some 750 to 800 species of acacias or 'mimosas' as they are affectionately known to the British. Many are Australian, either evergreen shrubs of various height or small to medium trees. *A. dealbata* is one of the best known to Europeans as it is widely grown in the south, making a tree of 15 m (50 ft) or more. **Leaves:** alternate, silvery, 8–13 cm (3–5 in) long, doubly pinnate with many narrow leaflets. **Flowers:** in large sprays, highly fragrant, individually like small, round, yellow, fluffy balls. *A. mearnsii,* from Tasmania, is very similar, also *A. decurrens*—both commonly cultivated as *A. dealbata*.
Season Winter and early spring.
Remarks Cultivated in France for perfumery purposes and as cut flowers. Bark yields a gum arabic substitute and is used for tanning. Most acacias have yellow flowers, either in round heads or spikes, and are good honey producers. Some kinds are thorny.

Adenium obesum   (*A. arabicum*)   Ψ occ. Ψ 🍃 ◁
Desert rose

Family *Apocynaceae*   **Place of origin** E. Africa
Description A fleshy stemmed evergreen shrub or small tree, up to 4·5 m (15 ft) with a swollen trunk. **Leaves:** stemless, oblong to egg shaped, up to 15 cm (6 in) long, clustered at ends of branches. **Flowers:** showy, in terminal clusters, rich rosy pink, paler towards centre, funnel-shaped, about 5 cm (2 in) across with 5 segments, downy outside.
Season Spring.
Remarks Only found in hot arid places. In very dry climates the plant often loses its foliage for much of the year. Juice of the plant contains a cardiac poison, used by Africans as an arrow poison.

Acacia mearnsii
Adenium obesum

Aleurites moluccana (*A. triloba*) Ψ⌀❋
Candlenut; candleberry tree; varnish tree; Indian walnut

Family *Euphorbiaceae* **Place of origin** Malaysia, Pacific Islands
Description A large evergreen tree with milky juice, up to 18 m
(60 ft) high, readily recognized by the frosted or scruffy appearance
of the young leaves, discernible even at a distance. **Leaves**: alternate,
simple, to 20 cm (8 in) long, ovate or spear-shaped or sometimes
three- or five-lobed low down on the tree. **Flowers**: in 18–23 cm
(7–9 in) clusters, small, yellowish, on whitish stems. **Fruits**: clusters
of green, 5 cm (2 in) fruits each containing 2 seeds.
Season Spring, summer.
Remarks Fruits purgative eaten raw, but this property disappears
with cooking. Valuable oils are extracted from the seeds, used in
soaps, paints and varnishes.

Annona muricata Ψ⌀֍
Soursop; prickly custard apple; guanábana

Family *Annonaceae* **Place of origin** South America
Description A small tree of the warm tropics, up to 6 m (20 ft).
Leaves: evergreen, alternate, leathery, glossy, entire, elliptic or
oblong, sharp-pointed, to 15 cm (6 in) long, rusty beneath. **Flowers**:
solitary, six-petalled, somewhat fleshy, 2·5 cm (1 in) long, yellow,
slightly fragrant. **Fruits**: large, ovoid, deep dark green covered with
curved fleshy spines, up to 20 cm (8 in) long. Some specimens weigh
3–4 kg (6–8 lbs).
Season Summer.
Remarks The largest fruited species of this genus; eaten raw or in
jellies and preserves, also made into sherbets and other drinks.

Annona reticulata Ψ⌀֍
Custard apple; bullock's heart; corazón

Family *Annonaceae* **Place of origin** Tropical America
Description Widely grown in the old world tropics as well as the
West Indies and parts of South America for its edible, although not
wildly exciting fruits. A deciduous small tree of 6–9 m (20–30 ft).
Leaves: alternate, entire, oblong-lanceolate, 18–20 cm (7–8 in) long,
pointed at tip. **Flowers**: clustered in the leaf axils, each with 6 petals,
the outer 3 fleshy, 2·5 cm (1 in) long, yellowish. **Fruits**: heart-shaped,
weighing from 0·2–2·25 kg ($\frac{1}{2}$–5 lb) according to variety, to 13 cm
(5 in) across. Reddish-yellow, becoming brown, with a netted
surface, flesh yellowish.
Season Summer.
Remarks Fruit rather flavourless.

Aleurites moluccana

Annona muricata

Annona reticulata

Araucaria heterophylla (A. excelsa) Ψ⌀
Norfolk island pine

Family *Araucariaceae* **Place of origin** Norfolk Island
Description Well known as a house plant, this handsome conifer
grows to a tremendous height under subtropical and tropical conditions. It will grow to 60 m (200 ft) in time, and then becomes leggy,
when the lower branches fall. It is much planted in the warmer parts
of America and neighbouring islands. **Leaves:** dark green, leathery,
stiff, awl-shaped, closely overlapping. Male and female cones on
same tree.
Season All year round; young plants are the most attractive.

Artocarpus altilis (A. incisus) Ψ⌀
Breadfruit

Family *Moraceae* **Place of origin** Malaya
Description An important tropical evergreen tree grown for its
large and edible fruits. Height ultimately around 12–18 m (40–60 ft).
Leaves: handsome, alternate, entire, dark glossy green, leathery,
ovate, pinnately lobed, around 60 cm (2 ft) long. **Flowers:** male and
female distinct, males yellow in stiff 15–30 cm (6–12 in) catkin-like
spikes, females forming round heads. **Fruits:** round or oval, prickly,
15–20 cm (6–8 in) across—look like large melons.
Season Summer.
Remarks Fruits eaten boiled or fried. They have a high starch and
protein content so are very nutritious. The bark is very tough and
when beaten out makes a native cloth.

Bauhinia purpurea Ψ occ. Ψ⌀☾
Orchid tree; mountain ebony; pink camel's foot; ebony wood

Family *Leguminosae* **Subfamily** *Caesalpiniodeae*
Place of origin China, India
Description *B. variegata* is a beautiful small tree, deciduous (or
evergreen in hot, moist areas) growing 4·5–9 m (15–30 ft), with a dark
brown, fairly smooth trunk. **Leaves:** shaped like the footprint of a
camel, alternate, two-lobed, 10–15 cm (4–6 in) across. **Flowers:**
orchid-like in short sprays or occasionally axillary or at ends of
branches, 7 cm (2½ in) across, rich pink or sometimes magenta or
mauve with crimson markings, 5 petals, 5 arched stamens, richly
fragrant. There is a white form with yellow markings. *B. purpurea* is
similar and often mistaken for *B. variegata* but flowers earlier and
does not have a white form.
Season Winter.
Remarks Pods, leaves and buds used as a vegetable in India. Wood
heavy and hard, used for agricultural implements, bark for tanning.

Araucaria heterophylla

Bauhinia purpurea

Artocarpus altilis

Bixa orellana ♈⌀✳

Lipstick tree; annatto; achiote

Family *Bixaceae*   **Place of origin** W. Indies, tropical America
Description Fast growing evergreen tree to about 6 m (20 ft),
planted in many parts of the tropics. **Leaves:** alternate, entire, thin,
oval heart-shaped, about 15 cm (6 in) long. **Flowers:** in panicles,
two- to four-flowered, pale pink, rose or white, full of yellow stamens, 5 cm (2 in) across. Seed capsules 5 cm (2 in) across, reddish or
crimson, covered with soft bristles, seeds orange-red.
Season Spring.
Remarks Boiled seeds or their coverings yield yellow and red dyes
used to colour cheeses and butter, also rice, soaps, etc. Various South
American Indians paint their bodies with a red paste obtained from
the seeds, both for ornament and as an insect repellent.

Blighia sapida (*Cupania sapida*) ♈‡✳

Akee

Family *Sapindaceae*   **Place of origin** West Africa
Description An evergreen tree of 9–12 m (30–40 ft). **Leaves:**
alternate, pinnate, with 6–10 oblong leaflets, 13–15 cm (5–6 in) long.
Flowers: in upright axillary racemes, white, small. **Fruit:** three-
angled, about 8 cm (3 in) long, red, splitting to reveal shiny, black
seeds, each with a white, fleshy aril.
Season Summer and autumn.
Remarks Named after William Bligh, captain of the 'Bounty'. Ripe
arils of fruits are edible and popular in West Indies when cooked
with fish. The seeds are poisonous, as are the unripe arils.

Bombax ceiba (*B. malabaricum*) ♈✳⌐0

Red silk cotton tree

Family *Bombacaceae*   **Place of origin** Tropical Asia
Description Large deciduous tree, eventually to 22·5 m (75 ft).
Trunk spiny, especially when young. **Leaves:** alternate, palmate
with 3 to 7 leaflets up to 25 cm (10 in) long. **Flowers:** solitary,
towards end of branches, large, five-petalled, crimson-scarlet
(occasionally white or purplish), 10 cm (4 in) long, full of scarlet
stamens tipped with purple. **Fruit:** a pod, about 15 cm (6 in) long
filled with silky cotton enclosing the seeds.
Season Spring, flowers appearing in quantity before the leaves.
Remarks Fleshy calyces behind flowers are eaten by Burmese in
curries. Cotton from pods used for stuffing pillows in India.

Bixa orellana
Bombax ceiba

Blighia sapida

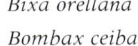

Brachychiton acerifolius (*Sterculia acerifolia*) Ψ⌀✳

Flame tree

Family *Sterculiaceae* **Place of origin** New South Wales, Australia
Description A deciduous or sometimes semi-deciduous tree up to
18 m (60 ft) tall. **Leaves:** long stemmed, maple-like, deeply lobed
into 5 or 7 leaflets, up to 25 cm (10 in) across. **Flowers:** showy, on
the bare branches, in large loose sprays or panicles, each bloom
about 2 cm (¾ in) long, bell-shaped, rich red with 5 sepals. There are
no petals.
Season Spring.
Remarks The inner bark is used to make hats, mats and baskets.

Brassaia actinophylla (*Schefflera actinophylla*) Ψ✳◇

Octopus tree; Queensland umbrella tree; Australian umbrella tree

Family *Araliaceae* **Place of origin** Australia
Description A handsome evergreen frequently cultivated as a
house plant in Europe and North America. In the subtropics it makes
a many-trunked tree up to 10 m (33 ft). **Leaves:** light green and glossy,
spread like a wheel, with 7–16 oblong leaflets up to 30 cm (12 in)
long, radiating from a central point. **Flowers:** crowded in terminal
inflorescences on radiating spikes, like the tentacles of an octopus,
rich red.
Season Early spring.
Remarks Common ornamental in Hawaii, California, Australia and
the Pacific islands.

Carica papaya Ψ✳✳

Papaya; pawpaw; tree melon

Family *Caricaceae* **Place of origin** Tropical America
Description Widely grown tropical fruit. Succulent, tree-like
plant, 4·5–6 m (15–20 ft) tall with a stout upright, leafy stem and
milky juice. **Leaves:** near top of stem, palmate, with seven lobes,
the latter pinnately divided, and up to 60 cm (2 ft) across. **Flowers:**
male and female usually on separate plants but sometimes herm-
aphrodite, sweet smelling, white; the male long-stemmed on 60–
90 cm (2–3 ft) axillary racemes, female on short stems and fewer.
Fruits: large, oval to round, yellowish when ripe, to 45 cm (1½ ft)
long, and weighing 0·9 kg (2 lb)—sometimes much more; smooth
skinned, fleshy, juicy, seeds black.
Season All year round.
Remarks Flesh of fruits delicious and popular in tropics, eaten raw
or preserved. Juice and leaves contain papain, a meat tenderizer.
Unripe fruits can be eaten as salad, or cooked as a vegetable. Also
made into chutney and a rum-type beverage.

Brachychiton acerifolius
Brassaia actinophylla

Carica papaya

Cassia fistula   ♆‡♎

Golden shower; Indian laburnum; pudding pipe tree; golden rain; purging cassia

Family *Leguminosae*   **Subfamily** *Caesalpinioideae*
Place of origin India
Description Beautiful deciduous tree to 12–15 m (40–50 ft). **Leaves:** alternate, smooth, 30–38 cm (12–15 in) long consisting of 3 to 8 pairs of 15 cm (6 in), ovate leaflets, downy and pink tinged when young. **Flowers:** clear yellow, 5 petals of unequal size, with thread-like, curling stamens, fragrant, in pendent 30–45 cm (1–1½ft) sprays. **Fruits:** long brown, cylindrical pods up to 90 cm (3 ft) long full of shiny brown seeds.
Season Spring.
Remarks Pulp from pods used to flavour tobacco in India. Bark used for tanning, roots as a laxative.

Cassia multijuga   ♆‡♎

Senna

Family *Leguminosae*   **Subfamily** *Caesalpinioideae*
Place of origin Guyana, Brazil
Description A short-lived but handsome small tree up to 6 m (20 ft) tall, very useful in small gardens, being quick growing and flowering in two years, from either cuttings or seed. **Leaves:** alternate, evenly pinnate with 18–30 pairs of leaflets, each about 2 cm (¾in) long, and linear oblong. **Flowers:** in large clusters on erect, terminal, 15 cm (6 in) panicles, bright yellow, each 5 cm (2 in) across with 5 petal-like divisions and 10 stamens. **Fruits:** flat pods 10–15 cm (4–6 in) long.
Season Spring.
Remarks Naturalized in parts of West Indies.

Casuarina equisetifolia   ♆

She oak; Australian pine; beefwood; horsetail tree; south sea ironwood

Family *Casuarinaceae*   **Place of origin** Australia
Description Easily recognizable evergreen trees on account of their long, slender, drooping branches. There are a number of species of which this is perhaps the most widely cultivated, usually as a windbreak near the sea. It is commonly seen in the Mediterranean region, sometimes kept clipped for hedging purposes. Height up to 21 m (70 ft), with slender, jointed branchlets. **Leaves:** reduced to mere scales. **Flowers:** insignificant. **Fruits:** round or oval cones.
Season Attractive all year round.
Remarks They make good timber trees, the hard wood taking a fine polish.

Cassia fistula
Cassia multijuga

Casuarina equisetifolia

Cecropia angustifolia (*C. digitata*)

Family *Urticaceae* (*Moraceae* by some authorities)
Place of origin Andes
Description Cecropias are easily recognized by their large, hand-shaped leaves and generally silvery foliage which can be spotted at a considerable distance. In the wild the hollow stems are inhabited by fierce ants which rush out and attack any animal, person or creature touching the trees. Most cecropias grow between 12–18 m (40–60 ft).
Leaves: evergreen, large, palmately lobed, entire, silvery beneath, long-stalked. **Flowers:** uninteresting, yellow spikes. Sap milky.
Season Attractive all year round.
Remarks The hollow stems of *C. peltata* are used by Indians to convey messages, hence its common name—trumpet tree.

Cercis siliquastrum

Judas tree; love tree

Family *Leguminosae*　**Subfamily** *Caesalpinioideae*
Place of origin S. Europe, W. Asia
Description A common tree of the Mediterranean region, spectacular in spring when the leafless branches are festooned with clusters of purplish rose, pea-shaped flowers each about 2 cm ($\frac{3}{4}$ in) in length. These are often borne directly on the branches or even on old parts of the trunk. Height up to 12 m (40 ft). **Leaves:** alternate, entire, roundish at base, as broad as they are long, about 10 cm (4 in), smooth. **Flowers:** in clusters of 3 to 6 but in great quantity. **Fruits:** flat pods, 8–10 cm (3–4 in) long, hanging in bunches.
Season Spring.

Chorisia speciosa

Floss silk tree

Family *Bombacaceae*　**Place of origin** Brazil, Argentina
Description A widely cultivated tree in the tropics and subtropics. Height 15 m (50 ft) or more, trunk enlarged at base and studded with stout, sharp thorns. **Leaves:** alternate on stems, palmate with 5 to 7 leaflets, all lanceolate, up to 13 cm (5 in) long with toothed edges and sharp points. **Flowers:** very showy, 8–13 cm (3–5 in) across, variable in colour, pink or white, usually spotted, five-petalled with whorls of staminodes, downy, normally appearing before the leaves. **Fruits:** about 20 cm (8 in) long.
Season Autumn, early winter.
Remarks The silky floss surrounding the seeds is used for stuffing pillows and cushions.

Cecropia angustifolia

Cercis siliquastrum

Chorisia speciosa

Cochlospermum vitifolium (*Maximiliana vitifolium*)
Wild cotton

Ψ ✿ ⚇

Family *Cochlospermaceae* **Place of origin** Venezuela, Central America, Mexico
Description A slender deciduous tree with reddish-brown branches which form a flat crown. Height from 4·5–12 m (15–40 ft) according to age and situation. **Leaves:** alternate, long stemmed, palmate, 10–30 cm (4–12 in) wide with 5 deeply cut lobes with pointed tips.
Flowers: spectacular, occurring when trees are leafless, in terminal clusters, brilliant yellow, 10–12 cm (4–4½ in) across, looking like huge buttercups, velvety with numerous scarlet-streaked orange stamens.
Fruits: pear-shaped, 8 cm (3 in) long, seeds embedded in long white floss.
Season Winter and early spring.
Remarks Floss from seeds used to stuff pillows and to provide cotton.

Cocos nucifera
Coconut palm

Ψ ‡ ✿

Family *Palmae* **Place of origin** Possibly the Cocos and Keeling Islands
Description Coconut palms are found throughout the tropics, usually along or near coastlines. Trees grow to about 24 m (80 ft), often with curved trunks, thickened at the base. **Leaves:** form a crown at top of the trunk, individually up to 6 m (20 ft) long, pinnate with 90 cm (3 ft) segments, leathery and glossy. **Flowers:** in large inflorescences up to 90 cm (3 ft) long, male and female flowers separated, former fragrant. **Fruit:** large, to 30 cm (1 ft), nut enclosed in thick fibrous husk. These take 10 or 12 months to ripen.
Season All year round.
Remarks A most valuable tropical plant. The liquid makes a refreshing drink; leaves used for thatching and mat making, trunk for cabinet work; nuts for confectionery, margarine, oil and soaps.

Cordia sebestena
Scarlet cordia; aloe wood; geiger tree

Ψ ⌀ ⌐o

Family *Boraginaceae* **Place of origin** Cuba, West Indies, Florida
Description An evergreen shrub or small tree widely cultivated in the tropics. Height 4·5–7·5 m (15–25 ft). **Leaves:** alternate, simple, ovate, 10–15 cm (4–6 in) long, 5–8 cm (2–3 in) wide, rough to the touch, sometimes toothed. **Flowers:** open, clustered at ends of branches, base tubular expanding to five- to eight-lobed bloom, up to 5 cm (2 in) across, scarlet or orange. **Fruits:** white, nut-like, about 2 cm (¾ in) long.
Season Summer, but often appears in other seasons.

Cochlospermum vitifolium

Cocos nucifera

Cordia sebestena

Cordyline australis (*Dracaena australis*) Ψ⌀✳

Cabbage tree; palm lily; grass palm; fountain dracaena

Family *Agavaceae* **Place of origin** New Zealand
Description A striking tree-like plant, up to 12 m (40 ft) but usually seen much smaller. The long, bare stems are crowned with bushy heads of stiff, sword-shaped leaves, each up to 90 cm (3 ft) long and around 8 cm (3 in) wide. **Flowers:** white and extremely fragrant, in large crowded panicles 30–60 cm (1–2 ft) across. Individually these have 6 petal-like segments and 6 stamens. **Fruits:** white berries.
Season Spring to early summer.
Remarks Early New Zealand settlers used tender young heads of foliage in place of cabbage.

Couroupita guianensis Ψ⌀❀

Cannonball tree

Family *Lecythidaceae* **Place of origin** Guyana, Surinam
Description A tall, soft-wooded, tropical tree to 15–24 m (50–80 ft).
Leaves: often in rosettes at tips of branches, alternate, entire, oblong ovate, 15–30 cm (6–12 in) long, hairy along the veins on the undersides. **Flowers:** in racemes, up to 90 cm (3 ft) long, directly from the branches or from the trunk, each bloom 10–15 cm (4–6 in) across, with a fruity fragrance and looking like a fat yellow cushion with 6 red petals and many stamens on a curved central column. **Fruit:** large, round like a cannonball, reddish brown, up to 20 cm (8 in) across, on long stringy stems they have a most evil smell.
Season Spring and summer; flowers and fruit on the same tree.
Remarks The soft tree wood is used by Indians to treat skin diseases of livestock, and the yellow pulp from the fruits is eaten by coastal people.

Delonix regia (*Poinciana regia*) Ψ⚘☾

Royal poinciana; flamboyant; peacock flower, flame tree (Trinidad)

Family *Leguminosae* **Subfamily** *Caesalpinioideae*
Place of origin Madagascar
Description Handsome, wide-branching deciduous tree, much planted in tropics as a street tree. Height to 12 m (40 ft), umbrella-shaped top, no spines. **Leaves:** alternate, pinnate with up to 20 pairs of 10 cm (4 in) pinnae and many leaflets, each leaf up to 60 cm (2 ft) long, and lacy. **Flowers:** often before leaves, orchid-like, brilliant scarlet, 8–13 cm (3–5 in) across, 5 crinkly-edged petals one of which is patterned with red, yellow and white; red stamens, sepals red with green linings, many blooms in racemes at ends of branches.
Fruits: hard, flat, black pods 30–60 cm (1–2 ft) long which hang on for months.
Season Summer, odd blooms at other seasons.

Cordyline australis

Delonix regia

Couroupita guianensis

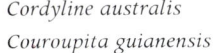

Dracaena draco   Ψ⌀

Dragon tree; dragon blood tree

Family *Agavaceae*   **Place of origin** Canary Islands
Description A curious tree which lives to a great age, one specimen
blown down in 1865 in Teneriffe was reputed to be over a thousand
years old. A mature specimen may be 18–21 m (60–70 ft) high with a
rounded umbrella like top. **Leaves:** perennial, in a crowded head,
individually sword-shaped, about 60 cm (2 ft) long and nearly 5 cm
(2 in) wide, glaucous green, entire. **Flowers:** insignificant, green,
very small in a panicle. **Fruits:** orange berries.
Season Grown as a curiosity and interesting at all seasons.
Remarks The dried sap—called dragon's blood—is used medicinally in plasters and for colouring varnishes.

Embothrium coccineum   Ψ⌀◇

Chilean fire tree; Chilean fire bush

Family *Proteaceae*   **Place of origin** Chile, Argentina
Description An evergreen tree from southern South America,
occasionally seen in temperate gardens when conditions are very
favourable. It does better in the subtropics than the tropics. Height
up to 12 m (40 ft), but also flowers as a shrub; suckering shoots.
Leaves: alternate, entire, leathery, glossy, ovate lanceolate, 5–8 cm
(2–3 in) long, dark green, paler underneath. **Flowers:** brilliant
scarlet and very profuse, in crowded axillary and terminal heads,
each about 5 cm (2 in) long, tubular with 4 recurved petal-like lobes
and long protruding stamens.
Season Summer.

Erythrina abyssinica *(E. tomentosa)*   Ψ♣◇

Red hot poker tree; karat tree

Family *Leguminosae*   **Subfamily** *Faboideae*
Place of origin Ethiopia
Description Several species are grown in the tropics, often as
deciduous street trees. The red hot poker tree has more rounded
inflorescences than most so is easily recognizable. Height about 12 m
(40 ft) with dark corky bark. **Leaves:** alternate, large, 3 leaflets, end
one the largest, orbicular, about 18 cm (7 in) long, 20 cm (8 in) wide.
Flowers: coral red, in dense rounded heads, individually nearly
5 cm (2 in) long, tubular with 5 slender lobes. **Fruits:** flat pods full of
red seeds.
Season Spring, before the leaves.
Remarks Seeds made into necklaces.

Dracaena draco

Embothrium coccineum

Erythrina abyssinica

Erythrina coralloides

Coral tree

Family *Leguminosae* **Subfamily** *Faboideae*
Place of origin Arizona, Mexico
Description All erythrinas have spectacular flowers, usually produced before the leaves. Widely planted as ornamentals in the tropics. *E. coralloides* grows to 6 m (20 ft). **Leaves:** prickly, alternate, 3 leaflets, triangular in shape, terminal ones about 11 cm (4½ in) long. **Flowers:** before the leaves, in dense, upright heads at tips of branches, bright scarlet, brownish buds, each bloom about 8 cm (3 in) long, narrow, pea-like. **Fruits:** pod 15–18 cm (6–7 in) long. **Season** Spring.

Erythrina crista-galli

Cockspur coral tree; cry-baby tree

Family *Leguminosae* **Subfamily** *Faboideae*
Place of origin Brazil, Argentina
Description Deciduous tree, up to 9 m (30 ft tall. Spectacular flowers which appear with the foliage. **Leaves:** alternate, leathery, with 3 oval, glaucous green leaflets, 8–10 cm (3–4 in) long. Vicious thorns are borne along the backs of the midribs. **Flowers:** pea-shaped, brilliant scarlet, each 4 cm (1½ in) long, 2·5 cm (1 in) wide, crowded in large terminal racemes. **Fruit:** a woody pod up to 30 cm (12 in) long, containing black and brown seeds.
Season Summer, or autumn if flowering on annual shoots.

Eucalyptus

Gum trees

Family *Myrtaceae* **Place of origin** Australia
Description There are some 500 species of eucalyptus, mostly Australian. Planted all over the tropics and subtropics, they are evergreen trees of graceful habit, some exceedingly tall, over 90 m (300 ft), others shrubby. **Leaves:** entire, opposite when young, alternate when adult; they may also change shape and width with maturity becoming tough, glaucous and oil-bearing. **Flowers:** conspicuous for their dense masses of stamens, borne singly or in umbels, white, pale yellow or scarlet. The coral gum, *E. torquata,* one of the brightest, grows about 9 m (30 ft) with 5 cm (2 in), pinkish-red flowers in terminal sprays and 4 cm (1½ in) urn-shaped fruits.
Season Spring and summer.
Remarks Eucalyptus have important economic uses, yielding timber and medicinal oils.

Erythrina coralloides
Eucalyptus torquata

Erythrina crista-galli

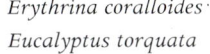

Feijoa sellowiana Ψ⌀ 𝔛

Pineapple guava; feijoa

Family *Myrtaceae* **Place of origin** South America, particularly Brazil.

Description An evergreen shrub or small tree, with whippy, white felted shoots, up to 6 m (20 ft) high. **Leaves:** opposite, simple, glossy green above, white, woolly beneath, oblong, 4–8 cm (1½–3 in) long, rounded at base, blunt-tipped. **Flowers:** solitary, in leaf axils, 4 cm (1½ in) across with 4 felted sepals, 4 fleshy petals which are white downy outside, purplish within, enclosing many crimson stamens. **Fruit:** an ovoid berry, pineapple-scented, 5–8 cm (2–3 in) long, green tinged with red, the seeds surrounded by a jelly-like pulp.

Season Summer.

Remarks Cultivated in tropical and subtropical regions for its fruits which are eaten raw or preserved as jam, jelly, or crystallized.

Ficus benghalensis *(F. indica)* Ψ⌀

Banyan; East Indian fig tree; Indian banyan

Family *Moraceae* **Place of origin** India, Burma

Description One of the world's largest tropical trees, epiphytic when young but then spreading outwards—sometimes over several acres—by putting down aerial roots which descend like ropes from the branches and enter the ground. Height to 30 m (100 ft). **Leaves:** evergreen, alternate, entire, thick and leathery, oval elliptic, about 20 cm (8 in) long. **Flowers:** growing inside small 1·2 cm (½ in) figs which appear in pairs.

Season All year round.

Remarks Tree held sacred by Hindoos who believe Brahma was transformed into a banyan tree. One specimen in India grew 600 m (2000 ft) across and was reputedly able to shelter 7000 people.

Ficus retusa Ψ⌀

Chinese banyan; Indian laurel; Malay banyan

Family *Moraceae* **Place of origin** India, Ceylon, Malaysia, China, New Caledonia

Description A large tree up to 20 m (60 ft) high with a wide spreading crown. It is a strangler fig, starting as an epiphyte and then producing very prominent surface roots as well as a few aerial roots which in time grow downwards from the branches and make new trunks. **Leaves:** broadly ovate with rounded base, blunt-tipped, fleshy, about 10 cm (4 in) long. **Fruits:** sessile, in pairs in the leaf axils, small, 8 mm (⅓ in) across, purplish.

Season Attractive all year round.

Feijoa sellowiana
Ficus benghalensis

Ficus retusa

Jacaranda mimosifolia *(J. mimosaefolia; J. ovalifolia)*

Fern tree (in West Indies); green ebony (U.S.A.)   Ψ ♇ ✳

Family *Bignoniaceae*   **Place of origin** Argentina
Description Shrubs and trees all native to tropical America. Much planted as street trees in tropics and subtropics; easily recognized since all the 50 odd species have fern-like foliage and bell-shaped, blue or violet flowers. This species grows to 15 m (15 ft) or more, with slender, smooth trunk. **Leaves:** deciduous, opposite, bipinnate with many leaflets, whole leaf about 45 cm (1½ ft) long. **Flowers:** bluish mauve, in axillary or terminal clusters, funnel-shaped, pendent, about 5 cm (2 in) long, 1·2 cm (1 in) wide at mouth with 5 turned back lobes with fine hairs inside throat.
Season Spring to early summer.

Kigelia pinnata   Ψ ‡ ⌂

Sausage tree; cucumber tree

Family *Bignoniaceae*   **Place of origin** Tropical Africa
Description Grown as a curiosity in tropical and subtropical gardens. Evergreen tree, to 15 m (50 ft), with a trunk of 60–75 cm (2–2½ ft) diameter, and light, flaking bark. **Leaves:** pinnate, whorled, up to 60 cm (2 ft) long, with opposite leaflets, normally around 6 cm (2½ in) long, terminal leaflet smaller. **Flowers:** dark reddish purple, campanulate, 8 cm (3 in) across on long stems, pollinated by bats. **Fruits:** large, sausage-shaped, up to 1 m (3 ft) in length with many seeds, swinging on long cord-like stems. They sometimes weigh over 7 kg (15 lb) each. Unripe fruits poisonous.
Season Autumn to spring.

Mangifera indica   Ψ⌀✳

Mango

Family *Anacardiaceae*   **Place of origin** India, Malaya
Description Densely foliaged tree of 20 m (60 ft) or more, wide-spreading and valuable for the shade it provides as well as its fruit. **Leaves:** abundant, alternate, leathery, deep glossy green, lanceolate, to 30 cm (12 in), pointed at tip, reddish when young and fragrant when bruised. **Flowers:** small, yellowish, in large panicles, fragrant. **Fruits:** also fragrant, variable in size, texture and colour, but frequently yellowish-red, oval to heart-shaped, about 8–13 cm (3–5 in) long and smooth skinned. Always with one large flat seed.
Season Winter, early spring but variable.
Remarks An important tropical fruit; used as a dessert and in chutneys.

Jacaranda mimosifolia

Mangifera indica

Kigelia pinnata

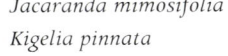

Melia azedarach (*M. australis; M. japonica*) Ψ‡✳ & ▭

Chinaberry; bead tree; Persian lilac; Chinatree; pride of India;
paradise tree

Family *Meliaceae* **Place of origin** N. India, China
Description A fast growing, widely cultivated, deciduous tree,
wide spreading to 12 m (40 ft); often used for street planting. Bark
dark brown with deep furrows. **Leaves:** 30–60 cm (1–2 ft) long,
bipinnate with ovate, toothed, pointed leaflets 2·5–5 cm (1–2 in)
long. **Flowers:** small, loosely clustered on short-stemmed 10–20 cm
(4–8 in) sprays, fragrant, mauve or lilac petals surrounding a deep
purple tube holding the stamens. **Fruit:** yellow, berry-like, 6–12 mm
($\frac{1}{4}$–$\frac{1}{2}$ in) long, staying long on the trees.
Season Spring.
Remarks Seeds made into beads and rosaries; timber used for furniture, oil obtained from roots.

Parkinsonia aculeata Ψ✦✳ & ▭

Jerusalem thorn; palo verde

Family *Leguminosae* **Subfamily** *Caesalpinioideae*
Place of origin Tropical America
Description Widely grown, dainty, spiny shrub or small tree, with
slender, frequently drooping, green branches. **Leaves:** alternate,
bipinnate, to 40 cm (16 in) with tiny, linear leaflets. **Flowers:** in
racemes, fragrant, bright yellow with 5 spreading petals and tubular
orange centre, each 1·5 cm ($\frac{2}{3}$ in) long. **Fruits:** narrow pods.
Season Spring.

Persea americana (*P. gratissima*) Ψ⌀✳

Avocado pear; alligator pear; aguacate

Family *Lauraceae* **Place of origin** Tropical America
Description Tropical evergreen, much-branched tree, growing
12–15 m (40–50 ft) high; in the Galápagos Islands we have seen them
over 18 m (60 ft). **Leaves:** opposite, entire, dark green, oval egg-
shaped, 10–20 cm (4–8 in) long with pointed tips. **Flowers:** small,
greenish, about 12 mm ($\frac{1}{2}$ in) across, in small terminal clusters. **Fruits:**
up to 15 cm (6 in) long, pear-shaped with thick glossy skins, green to
brownish-purple with one large seed, flesh solid, buttery, cream-
coloured. Fruits usually hang on long stems.
Season Summer, but variable.
Remarks Avocados are highly nutritious and contain much oil and
many vitamins.

Melia azedarach
Parkinsonia aculeata

Persea americana

Plumeria rubra

Frangipani; temple tree; pagoda tree

Family *Apocynaceae*   **Place of origin** Mexico to Panama
Description Popular and widely grown all over the tropics and subtropics. Large deciduous shrub or small tree to around 7·5 m (25 ft). Thick, brittle branches which look gaunt and ugly when bare of foliage. **Leaves:** alternate, simple, oblong-lanceolate, large, to 20 cm (8 in) long on 8–13 cm (3–5 in) stems. **Flowers:** showy, in terminal clusters, 5 overlapping petals, pink or red in *P. rubra*; white with a yellow centre in the form known as *acuminata*. There are also yellow forms. All are exceedingly fragrant.
Season Summer to winter.
Remarks In the 12th century an Italian nobleman called Frangipani concocted an elusive scent which became the rage of Europe. When plumerias were discovered, the fragrance was found to be identical —hence its name.

Psidium guajava

Guava; apple guava; common guava; lemon guava; guayabillo

Family *Myrtaceae*   **Place of origin** Tropical America
Description Shrub or small tree to 7·5 m (25 ft), with scaly, greenish bark which often peels off in strips. **Leaves:** opposite, simple, ovate to oblong-elliptic, with prominent veins, green, downy beneath. **Flowers:** white, about 2·5 cm (1 in) across, five-petalled, solitary or several together, fragrant. **Fruit:** pear-shaped, yellow, 2·5–10 cm (1–4 in) long, aromatic when ripe, flesh yellow or pink.
Season Spring, summer.
Remarks Edible fruits used for preserves.

Ravenala madagascariensis

Traveller's tree; traveller's palm

Family *Strelitziaceae*   **Place of origin** Madagascar
Description A strange looking plant of small tree-like proportions. From short, palm-like trunks formed from the bases of old leaf stalks, emerge large, 2·4–3 m (8–10 ft) banana-like leaves on long stems. They are palm-like and spread in a half circle up to 9–10·5 m (30–35 ft). **Flowers:** in axillary racemes, with large 30 cm (12 in) stiff, white bracts, each having 6 stamens. **Seeds:** round, pea-sized, black with a bright blue patch over half their surface, edible.
Season All year foliage interest, flowering in spring.
Remarks Bases of leaf-stalks and bracts hold water—hence the common name.

Plumeria rubra

Psidium guajava

Ravenala madagascariensis

Roystonea regia (*Oreodoxa regia*) Ψ‡❋
Royal palm

Family *Palmae*   **Place of origin** Cuba
Description There are many types of palm in the tropics and subtropics, but this is one of the most distinctive. Lofty, to a height of 12–21 m (40–70 ft), it has a greyish-brown thickened trunk base, bulges again towards the middle and then tapers upwards. **Leaves:** forming a crown at summit of tree, fronds pinnate, leaflets set in 4 rows, arching, each frond up to 3 m (10 ft) long. **Flowers:** male and female separate, individually small but borne in large clusters springing directly from the trunk just below the leaves. **Fruits:** small, purple in huge bunches, 23–27 kg (50–60 lb) in weight.
Season Attractive all year round.
Remarks Royal palms are often planted as an avenue leading up to an important building or residence.

Schinus molle Ψ‡❋
Pepper tree; Peruvian mastic tree; molle

Family *Anacardiaceae*   **Place of origin** Peru
Description Evergreen, of rounded shape but drooping branches, 9–15 m (30–50 ft) high. **Leaves:** alternate, pinnate, 13–23 cm (5–9 in) long with narrow, linear leaflets about 5 cm (2 in) long. These contain an oily principle and are aromatic when crushed. **Flowers:** very small, five-lobed, in short 2·5–4 cm (1–1½ in) panicles, yellow-white. **Fruits:** round, rosy-red, like small peas.
Season Summer.
Remarks Ground seeds used as a pepper adulterant in Mexico, also a drink. The ground bark is a purgative.

Spathodea africana (*S. nilotica*) Ψ occ. Ψ❋❋
African tulip tree; flame of the forest (Jamaica); red tulip (Trinidad); tulipán

Family *Bignoniaceae*   **Place of origin** Tropical Africa
Description Magnificent tree widely cultivated in the tropics for its brilliant flowers. Evergreen (although some specimens lose their leaves in winter), up to 21 m (70 ft). **Leaves:** opposite, compound, with odd number of leaflets, between 7 and 19, dark green. **Flowers:** in terminal racemes, up to 13 cm (5 in) long, tulip-shaped, brilliant scarlet, outlined with yellow. Brown, boat-shaped buds full of liquid.
Season Spring, but occasional blooms at other seasons.
Remarks In parts of Africa this tree is connected with magic and witchcraft; tribal drums are made from the wood, also witch-doctors' magic wands. Bark and flowers are used medicinally.

Roystonea regia   *Schinus molle*

Spathodea africana

Syzygium malaccense (S. jambos; Eugenia malaccensis)

Malay apple; pomerack jambos   Ψ⌀◇

Family *Myrtaceae*   **Place of origin** Malay peninsula
Description Attractive, small evergreen tree of 9–12 m (30–40 ft)
which sometimes drops its leaves during the flowering season.
Leaves: opposite, simple, short-stalked, elliptic, 20–30 cm (8–12 in)
long by half as wide; leathery. **Flowers:** showy, a few together in
axillary cymes, bright purplish-red, 4 to 5 petals with many stamens.
Fruits: red, pink or occasionally white, about 5 cm (2 in) long.
Season Summer.
Remarks Fruit eaten raw or cooked, also preserved and made into
wine.

Tabebuia rosea (T. pentaphylla)   Ψ❁◁

Pink poui; rosy trumpet tree

Family *Bignoniaceae*   **Place of origin** Central and South America
Description Common and showy flowering tree of the New World
tropics and subtropics, particularly where there are moist conditions.
Tall trees, 18–21 m (60–70 ft) or more with smooth, straight, slender
trunks which only branch near the tops. **Leaves:** deciduous,
opposite, palmate, usually with five-stalked, elliptic-oblong leaflets
of unequal size, the largest 15 cm (6 in) or more long and 6 cm (2½ in)
wide. **Flowers:** usually before the leaves, in dense panicles, each
over 8 cm (3 in) long, white to deep rose, funnel-shaped with a
yellow throat and crinkly edges; quickly fading.
Season Spring.
Remarks The timber, sold as amapa wood, being attractively marked
and patterned is used for furniture.

Tabebuia serratifolia (T. flavescens)   Ψ❁◁

Trumpet tree; yellow poui

Family *Bignoniaceae*   **Place of origin** Colombia to Bolivia
Description Large tropical tree, slow growing to 42 m (140 ft) or
even 60 m (200 ft); trunk slender, bark greyish. **Leaves:** opposite,
compound with 5 oblong or lanceolate leaflets, each up to 15 cm
(6 in) long. **Flowers:** appearing before the leaves, in dense heads,
clear yellow, individually funnel-shaped with 5 crinkly lobes, and
about 8 cm (3 in) long.
Season Winter or early spring.
Remarks Trees plentiful in Caribbean area, often planted for their
showy winter flowers. Wood tough, heavy, hard and strong, used
for railway sleepers, wharves, bridges, fences and cabinet making.
Locally known as *pao d'arco* or Surinam greenheart.

Syzygium malaccense

Tabebuia rosea

Tabebuia serratifolia

2
Shrubs

Acalypha wilkesiana (*A. tricolor*)   Ψ⌀

Match me if you can; beefsteak plant; Jacob's coat; copper leaf

Family *Euphorbiaceae*   **Place of origin** Pacific Islands
Description A handsome evergreen shrub of rounded habit, from
0·9–6 m (3–20 ft) high and as much across. It is widely grown as a
garden ornamental, or as a small pot plant in cool climates. **Leaves:**
alternate, simple, smooth, lavishly produced, elliptic or ovate,
13–20 cm (5–8 in) long, with toothed edges. Their colouring con-
stitutes the showy part of the plant, being brownish-green splashed
with rose-pink, crimson, pink or brown often bordered with various
colours. No two leaves are exactly alike. **Flowers:** uninteresting,
reddish, in 15–20 cm (6–8 in) slender spikes.
Season All year round.
Remarks Particularly common in West Indies where they are
sometimes used for hedging.

Aeonium arboreum 'Atropurpureum'   Ψ⌀✳

Family *Crassulaceae*   **Place of origin** Morocco.
Description A small perennial evergreen shrub 30–60 cm (1–2 ft)
tall, with woody stems, the few branches terminating in rosettes of
green, shiny leaves—the general effect reminiscent of giant semper-
vivums on rugged stems. **Leaves:** fleshy, lanceolate, 5–7·5 cm (2–3 in)
long by 2 cm ($\frac{3}{4}$in) wide. **Flowers:** crowded in dense 10 cm (4 in)
oval inflorescence, individually small, 1·5 cm ($\frac{3}{5}$in) across, yellow
with 9–12 petal segments.
Season Winter and spring.
Remarks Cultivars like 'Atropurpureum' and 'Foliis Purpureis'
have dark purple leaves. Portuguese fishermen use sap from the
leaves to toughen their fishing lines.

Acalypha wilkesiana
Aeonium arboreum 'Atropurpureum'

Aloe arborescens   ΨØ ▷

Candelabra aloe; tree aloe; torch plant

Family *Liliaceae*   **Place of origin** South Africa
Description A succulent shrub of 2·4–3·6 m (8–12 ft), suckering at base. **Leaves**: light green, thick, grooved, sharp-pointed with spiny margins, 60 cm (2 ft) long by 5 cm (2 in) wide, in rosettes at ends of the branches. These turn reddish in winter. **Flowers**: on thick 75 cm (2½ ft) stems which come from centre of rosettes, crowded at the tops of the spikes, individually tubular, 4 cm (1½ in) long, brick red.
Season Summer.
Remarks Most widely grown species of the S. African aloes.

Banksia   ΨØ ◇

Banksia; Australian honeysuckle tree

Family *Proteaceae*   **Place of origin** Australia
Description Trees or shrubs native to Australia but are now increasingly grown as ornamentals in subtropical gardens. There are about 50 species. **Leaves**: variously shaped, alternate, leathery, usually prickly and deeply toothed at margins, obovate to linear, paler underneath, 6 cm (2½ in) across. **Flowers**: in large, woody, cone-like spikes, without stalks, either terminal or axillary, very persistent. *Banksia ashbyi* has large, orange flower spikes, 15 cm (6 in) long and 9 cm (3½ in) wide on 3 m (10 ft) shrubs, with obovate, toothed leaves to 10 cm (4 in) long. *B. coccinea* grows to 4·5 m (15 ft) with scarlet, globular, cob-like spikes, 5 cm (2 in) across.
Season Spring to early summer, also autumn to winter.
Remarks There can be 1000 flowers in a single spike.

Brugmansia × candida *(Datura arborea; D. × candida)*

Angel's trumpet; maikoa   ΨØ ⊲

Family *Solanaceae*   **Place of origin** Species native to Andean regions, Colombia to Ecuador
Description A small tree or large shrub up to 3–6 m (10–20 ft) widely cultivated in the tropics and subtropics. **Leaves**: large, entire, oval oblong, hairy, around 20 cm (8 in) long. **Flowers**: pendulous, large, 25–30 cm (10–12 in), funnel-shaped flaring at mouth, corolla five-lobed, white, musky odour.
Season Summer.
Remarks Both seeds and leaves are narcotic. Flowers should be floated in water before arranging to prevent wilting. There is a double-flowered form usually known as *Datura cornigera* or *Brugmansia knightii*.

Aloe arborescens   *Banksia ashbyi*

Brugmansia × candida

Brugmansia sanguinea *(Datura sanguinea)*
Red angel's trumpet

Family *Solanaceae* **Place of origin** Colombia to Chile.
Description Shrub or small tree to 10·5 m (35 ft). **Leaves:** alternate, toothed when young but entire when mature, softly hairy, oval oblong with wavy margins, about 18 cm (7 in) long. **Flowers:** pendulous, tubular, 20–23 cm (8–9 in) long, red at mouth, otherwise overall colour yellow. Golden, orange and yellow-green forms are common.
Season Spring and summer.
Remarks All parts of the plant narcotic.

Brunfelsia pauciflora var. calycina *(B. calycina)*
Morning, noon and night; yesterday, today and tomorrow Ψ ∅ 🎱

Family *Solanaceae* **Place of origin** Brazil
Description Evergreen shrub of 1·2–1·8 m (4–6 ft). **Leaves:** alternate, simple, entire, oblong or oblong-lanceolate, rich green, leathery, 8–15 cm (3–6 in) long. **Flowers:** salver-shaped, with a long tube, calyx five-lobed, 5 petals, fragrant, deep purple fading with age to mauve and then almost white, each up to 5 cm (2 in) across. Borne in terminal cymes of 1 to 10.
Season All year round.

Caesalpinia pulcherrima *(Poinciana pulcherrima)*
Barbados pride; pride of Barbados; flower fence; dwarf poinciana; peacock flower; paradise flower Ψ ✿ 🎱

Family *Leguminosae* **Subfamily** *Caesalpinioideae*
Place of origin West Indies
Description A popular ornamental shrub much planted in tropical countries. Height up to 3 m (10 ft), rather prickly stems. **Leaves:** alternate, lacy looking, doubly pinnate with 3 to 9 pairs of pinnae which each has 5 to 10 pairs of 1·2 cm ($\frac{1}{2}$ in) oblong leaflets. **Flowers:** long stemmed in racemes or panicles, red with yellow borders and long bright red stamens to 6 cm ($2\frac{1}{2}$ in), followed by leathery pods. Flowers occasionally all yellow or rosy-red.
Season Autumn, spring, occasionally at other seasons.
Remarks Extract from pods and leaves used as a laxative in India and the roots to treat fevers in Angola.

Brugmansia sanguinea   *Caesalpinia pulcherrima*
Brunfelsia pauciflora var. *calycina*

Calliandra haematocephala (C. inaequilatera) Ψ ⚘ ◇
Powderpuff

Family *Leguminosae* **Subfamily** *Mimosoideae*
Place of origin Bolivia
Description Evergreen shrubs of loose habit widely grown for
their showy flowers which resemble fluffy powderpuffs. Plants
bloom when quite young and small, eventually reaching a height of
4·8 m (16 ft) or more. **Leaves:** alternate and doubly pinnate, carrying
5 to 10 pairs of leaflets, each oblong-lanceolate, 2–4 cm ($\frac{3}{4}$–$1\frac{1}{2}$ in)
long. **Flowers:** in round heads, 5–8 cm (2–3 in) across, composed
mostly of long, red-tipped, white stamens which gives them a very
light and dainty appearance.
Season Spring.
Remarks Other common powderpuffs are *C. tweedii* the Mexican
flamebud, rich red, to 1·8 m (6 ft); *C. portoricensis*, white, making a
large shrub or tree to 6 m (20 ft); and *C. fulgens,* pink with crimson
anthers, 3 m (10 ft).

Callistemon citrinus (C. lanceolatus)   Ψ⌀◇
Bottle brush ·

Family *Myrtaceae* **Place of origin** Australia
Description Callistemons are evergreen shrubs or small trees
widely planted in the subtropics, especially *C. citrinus*. Normally
about 3 m (10 ft) in cultivation, it grows double that height in the
wild. **Leaves:** alternate, simple, leathery, lanceolate with lateral
veins, 5–8 cm (2–3 in) long, with a faint lemon odour. **Flowers:**
showy, resembling bottle brushes, in 10 cm (4 in) spikes with a tuft
of leaves at the top of each stem, crimson with long, brilliant red
stamens. 'King's Park Special' is a particularly good form commonly
cultivated in gardens.
Season Spring.

Capparis spinosa (C. rupestris)   Ψ⌀⅋
Caper bush

Family *Capparidaceae* **Place of origin** Mediterranean region
Description A subtropical, spiny shrub, rather straggly with slender stems up to 1·5 m (5 ft) in height. **Leaves:** evergreen, alternate,
simple, rather fleshy, oval, to 5 cm (2 in), with 2 reflexed spines at
base of each leaf stalk. **Flowers:** solitary, 5–8 cm (2–3 in) across,
white on thick stems, 4 petals, 4 sepals. **Fruit:** a round or oval
berry.
Season Spring and summer.
Remarks The pickled flower buds are the capers of commerce.

Calliandra haematocephala   *Capparis spinosa*
Callistemon citrinus 'King's Park Special'

Cassia alata Ψ ‡ ♋

Candlestick senna; Christmas candle; empress candle plant; golden chain tree; ringworm cassia

Family *Leguminosae* **Place of origin** Tropical America
Description Evergreen shrub of 2–3 m (6–8 ft), the branches terminating in erect spikes of flowers. **Leaves:** alternate, pinnate, 45–75 cm (1½–2½ ft) long with 8 to 14 pairs of leaflets, each oblong or obovate, 6 cm (2½ in) long. **Flowers:** yellow in showy, crowded spikes. **Fruits:** angled pods with wings, about 15 cm (6 in) in length.
Season Late summer but occasionally at other seasons.
Remarks The leaves have been used in the treatment of ringworm.

Cestrum elegans (C. *purpureum*) Ψ⌀ ⊳

Purple cestrum

Family *Solanaceae* **Place of origin** Mexico
Description Large, spreading evergreen bush with downy pendulous shoots, normally 2·4–3 m (8–10 ft) tall. **Leaves:** alternate, entire, ovate or ovate-lanceolate, 8–13 cm (3–5 in) long, 2·5–5 cm (1–2 in) wide with pointed tips, pungent smelling. **Flowers:** nodding, tubular with 5 pointed lobes, in dense clusters, individual flowers crimson to reddish, 2·5 cm (1 in) long. **Fruit:** a round berry, red, 12 mm (½ in) wide. *C. newellii* is possibly a hybrid of this species.
Season Summer, autumn, occasionally at other seasons.

Cestrum nocturnum Ψ⌀ ⊳

Night blooming jessamine; galán de noche; hierba hedionda

Family *Solanaceae* **Place of origin** West Indies
Description An evergreen shrub up to 3·6 m (12 ft) high. Frequently planted close to houses on account of its scent. **Leaves:** alternate, entire, narrowly oval, lanceolate, 8–13 cm (3–5 in) long. **Flowers:** in long axillary sprays, only opening at night, each about 2·5 cm (1 in) long, narrowly tubular, green or whitish, extremely fragrant.
Fruits: small white berries.
Season Summer, autumn, spring in some areas.

Chamelaucium (Chamaelaucium) uncinatum

Geraldton wax Ψ⌀ ✿

Family *Myrtaceae* **Place of origin** Western Australia
Description Attractive, spreading shrub up to 3 m (10 ft) high and 4 m (13 ft) across with fibrous bark. **Leaves:** alternate, narrow, heath-like, dark green, about 2·5 cm (1 in) long, hooked at the tip.
Flowers: in terminal racemes, 5 stiff, waxy petals, 10 stamens to each rounded 12–20 mm (½–¾ in) flower, colours white, pink or red.
Season Spring to summer.

Cassia alata

Cestrum newellii

Cestrum nocturnum

Chamelaucium uncinatum

Clerodendrum speciosissimum (*C. fallax*)
Clerodendrum

Family *Verbenaceae* **Place of origin** Java
Description An erect shrub of 3–3·6 m (10–12 ft). **Leaves:** opposite, large, shining, ovate, pointed at tips, to 30 cm (1 ft) long, entire or toothed. **Flowers:** bright scarlet with protruding stamens, to 5 cm (2 in) long.
Season Spring, summer.
Remarks Often seen growing on walls and fences, suitable for sun or shade.

Codiaeum variegatum
Croton; variegated laurel

Family *Euphorbiaceae* **Place of origin** Malaysia, Pacific Islands
Description Evergreen shrubs grown all over the world for their brightly patterned leaves. In sunny tropical gardens they may reach 4·5 m (15 ft) in height; as pot plants in cool climates a mere 45 cm (1½ ft). Represented in cultivation by *C. v. pictum*; there are many forms and hybrids with variously shaped and coloured foliage. **Leaves:** alternate, simple, linear to lanceolate or oblong, margins entire or lobed, or cut almost to the midrib, often twisted or with appendages. Colours—green, yellow or red, spotted or suffused with white, red or yellow. **Flowers:** insignificant, small, whitish, in axillary racemes.
Season Attractive right through the year.
Remarks Croton oil is obtained from one species.

Congea tomentosa
Lavender wreath; shower orchid

Family *Verbenacea* **Place of origin** South East Asia
Description A climbing deciduous shrub, grown on trellises, fences and other supports. Height 1·8–3 m (6–10 ft), slender branches often drooping. **Leaves:** opposite, simple, elliptic-ovate, 10–20 cm (4–8 in) long, softly hairy beneath. **Flowers:** small, white, with 2 lipped petals, the one above divided into 2 lobes and the other into 3 lobes. Showy parts however are the large, long-lasting, 2·5 cm (1 in) bracts behind each bloom. These are downy, white to lilac and resemble flowers. Very profuse so general effect is quite spectacular.
Season Autumn, winter.
Remarks Flower sprays can be dried for winter arrangements.

Clerodendrum speciosissimum

Codiaeum variegatum

Congea tomentosa

Cyphomandra betacea

Ψ⊘ or ‡❋

Tree tomato; tomato tree; tamarillo; tomato deárbol

Family *Solanaceae*　**Place of origin**　South America
Description　A large, branching but spineless evergreen shrub with
downy stems and foliage, growing up to 3 m (10 ft). **Leaves:** alternate, either entire, rounded-ovate with 3 lobes or pinnately dissected,
softly hairy, unpleasant smelling when bruised, 20–25 cm (8–10 in)
long and 10 cm (4 in) wide. **Flowers:** in racemes, fragrant, five-lobed,
whitish pink with a dark stripe down the base of each segment.
Fruits: egg-shaped, about 5 cm (2 in) long, orange-red, smooth,
shiny, on long stems, edible.
Season　Spring through to autumn, continuously produced.
Remarks　Fruits eaten raw or cooked, also made into jam or jelly.
They are equally suitable for a fruit or a green salad.

Dillenia suffruticosa (*Wormia suffruticosa*)

Ψ⊘ 🆇

Family *Dilleniaceae*　**Place of origin**　Malaysia
Description　A large, heavily-branched, densely-leaved, shrub
normally 3–3·6 m (10–12 ft), although it will grow much higher.
Often planted on golf courses or in shrubberies, and kept compact
by annual clipping. **Leaves:** evergreen, alternate, large, simple,
elliptic to obvate, 23–25 cm (9–10 in), pinnately-veined. **Flowers:**
large, showy, bright yellow, solitary or clustered, five-petalled with
a fleshy, conical centre, 10 cm (4 in) across.
Season　Summer.
Remarks　The fleshy, star-like fruits of some species are edible and
used to make jams, jellies and curries in Asia.

Dracaena fragrans 'Massangeana' (*D. massangeana*)

Corn plant
Ψ⊘ ◇

Family *Agavaceae*　**Place of origin**　Guinea
Description　The type species *Dracaena fragrans* is a plant with
spreading, recurved leaves which will grow to 6 m (20 ft) in the
tropics. **Leaves:** basal, arching, 60–90 cm (2–3 ft) long, 10 cm (4 in)
wide, rich green. The form known as 'Massangeana' is similar but
lower growing, its leaves having a broad, sulphur-yellow stripe
down the centre of each. **Flowers:** fragrant, yellow, in long spikes,
each about 12 mm ($\frac{1}{2}$ in) long.
Season　Varies, according to age of plant.

Cyphomandra betacea

Dracaena fragrans 'Massangeana'

Dillenia suffruticosa

Eriobotrya japonica *(Photinia japonica)*

Loquat; Japanese medlar; Japanese plum

Family *Rosaceae*   **Place of origin** China
Description A large evergreen shrub around 3 m (10 ft) or small tree to 7·5 m (25 ft). **Leaves:** alternate, large, 15–25 cm (6–10 in) long by 8–13 cm (3–5 in) wide, firm, simple, with prominent veins, deep green, rusty downy beneath. **Flowers:** small, in stiff terminal panicles 8–15 cm (3–6 in) long, individually 6–18 mm ($\frac{1}{4}$–$\frac{3}{4}$ in), whitish, with 5 petals and 5 sepals, strongly scented. **Fruits:** ovoid, in loose clusters, apple-like, 2·5–4 cm (1–1$\frac{1}{2}$ in) long, orange-yellow, with hairy skin; flesh white or pale yellow, fragrant.
Season Late autumn and winter, fruits in spring.
Remarks Fruits eaten fresh, in drinks, jellies, candied or in fruit salads.

Euphorbia leucocephala

Pascuita; white lace euphorbia

Family *Euphorbiaceae*   **Place of origin** Mexico, Guatemala
Description A shrub or small tree, popular in the West Indies, with slender branches, and growing 1·5–3·6 m (5–12 ft). Stems contain milky sap which can produce dermatitis in susceptible individuals. **Leaves:** mostly in whorls, small, 5–8 cm (2–3 in) long, variously shaped from oblong to elliptic, to oblong lanceolate, stalks 5–8 cm (2–3 in) long. **Flowers:** in large inflorescences composed of a froth of small, white flowers in umbels; the general effect more like English cow parsley than a euphorbia.
Season Autumn to spring.

Euphorbia milii *(E. splendens)*

Crown of thorns; Christ thorn

Family *Euphorbiaceae*   **Place of origin** Madagascar
Description A succulent, very spiny shrub, growing 90–180 cm (3–6 ft) high, with brownish, branched stems carrying closely packed, thick, tapering spines. **Leaves:** few, carried near ends of the branches, alternate, obovate, entire, 6 cm (2$\frac{1}{2}$ in) long, stalked, thin, bright green. **Flowers:** bright scarlet in groups on sticky, branched, long-stemmed inflorescences, each 6–12 mm ($\frac{1}{4}$–$\frac{1}{2}$ in) across.
Season Spring, but intermittently most of the year, according to climate.

Eriobotrya japonica   *Euphorbia milii*

Euphorbia leucocephala

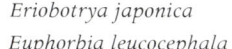

Euphorbia pulcherrima (*Poinsettia pulcherrima*)

Poinsettia; Christmas flower; Christmas star; Mexican flameleaf

Ψ⊘✳

Family *Euphorbiaceae* **Place of origin** Mexico
Description Well-known, winter-flowering shrub, widely grown for Christmas as a pot plant in cool climates. In the tropics and subtropics it grows to 2·4–3 m (8–10 ft), with short thick trunk and many brown-barked branches. **Leaves:** soft, entire, fiddle-shaped on ovate elliptic, toothed or lobed, 10–18 cm (4–7 in) long, bronzed when young. **Flowers:** small, green and yellow, in clusters set off by a circle of large scarlet bracts, 13 cm (5 in) long and half as wide. Cultivars with pink, white or shades of red bracts are common.
Season Winter.
Remarks Stems contain a milky sap which is poisonous to some people if it touches open wounds.

Galphimia glauca (*Thryallis glauca; G. nitida*) Ψ⊘◇

Family *Malpighiaceae* **Place of origin** Guatemala, Mexico
Description A shrubby plant often trained as a climber against fences. Height to 1·8 m (6 ft). **Leaves:** opposite, oblong or elliptic, simple, to 5 cm (2 in) long, short-stemmed. **Flowers:** yellow in dense, terminal racemes, each 2 cm ($\frac{3}{4}$ in) across.
Season Spring and summer.

Gardenia taitensis Ψ⊘⬡

Tahitian gardenia; tiare; tiare Tahiti; kiele

Family *Rubiaceae* **Place of origin** Tahiti and neighbouring South Pacific Islands.
Description Evergreen shrub of 4·5–6 m (15–18 ft), erect but with spreading branches. **Leaves:** opposite, large, dark green, glossy, ovate-lanceolate, to about 20 cm (8 in) long. **Flowers:** single, white, 6 petals, star-like, about 10 cm (4 in) across, exceedingly fragrant.
Fruits: berries.
Season In bloom practically all the year; easily detected—even at night—by its scent.
Remarks: Blossoms worn as leis, or in the hair on Pacific Islands; also macerated and used to alleviate headaches. In Hawaii often used for hedging. Does well in seaside localities.

Euphorbia pulcherrima

Galphimia glauca

Gardenia taitensis

Grevillea leucopteris ♉ ⚕ ◇
Spider flower

Family *Proteaceae*   **Place of origin** Western Australia
Description Shrub with long, leafless flowering stems to 3·5 m
(11½ft). **Leaves:** glaucous, fern-like, up to 25 cm (10 in) long. **Flowers:** in large cylindrical panicles, creamy white, fragrant, tubular
perianth with long style, no petals.
Season Spring and summer.

Hibiscus rosa-sinensis ♉ occ. ♉⌀ ✄
China rose; rose of China; shoe plant

Family *Malvaceae*   **Place of origin** China
Description Seen everywhere in the tropics. The type species is a
shrub or small tree of 2·4–4·5 m (8–15 ft). **Leaves:** alternate, smooth,
ovate, toothed, narrow at tip, to 15 cm (6 in) long. **Flowers:** showy,
bell-shaped, 10–13 cm (4–5 in) across, rich scarlet with a long central
column supporting stamens and stigma. Cultivars with larger flowers:
white, yellow, orange, pink and various shades of red, sometimes
double.
Season Spring, summer, autumn.

Hibiscus schizopetalus ♉ occ. ♉⌀ ✄
Japanese lantern

Family *Malvaceae*   **Place of origin** Tropical East Africa
Description A shrub of 1·8–2·4 m (6–8 ft) with slender drooping
branches. **Leaves:** alternate, oval elliptic, toothed margins, smooth,
10–13 cm (4–5 in) long. **Flowers:** pendent on long stems, orange-red,
streaked with pink, the petals deeply cut, with 5 cm (2 in) long
protruding stamen column; length of flower about 6 cm (2½in).
Season Summer, autumn.

Isoplexis canariensis (*Digitalis canariensis*) ♉⌀ ◇
Family *Scrophulariaceae*
Place of origin Canary Isles and Maderia
Description A bushy shrub, 1·5 m (5 ft) high, with erect, downy
shoots. **Leaves:** alternate, simple, evergreen, oval-lanceolate and
toothed, 13 cm (5 in) long, 2·5–5 cm (1–2 in) wide, downy beneath.
Flowers: many, in erect terminal spikes up to 30 cm (1 ft) in length,
two-lipped, upper longer than lower, orange-yellow, 2·5 cm (1 in)
across.
Season Summer.

Grevillea leucopteris
Hibiscus schizopetalus

Hibiscus rosa-sinensis
Isoplexis canariensis

Ixora fulgens

Flame of the woods; jungle geranium

Family *Rubiaceae* **Place of origin** Possibly Java or India, but uncertain
Description Handsome small shrubs from 1·2–3 m (4–10 ft) or more according to species. Often used for hedging or planted in borders in subtropics. **Leaves:** leathery, opposite or in whorls, lanceolate, to 13 cm (5 in), slender pointed, glossy and evergreen. **Flowers:** showy, in dense terminal heads, orange scarlet or coral red, individual corolla tubes 4–5 cm (1½–2 in) long, four-lobed, occasionally five. White, salmon, deep red and pale yellow flowers occur in some species.
Season Summer, autumn, spasmodically in spring.
Remarks The flowers and bark of ixoras provide lotions to soothe bloodshot eyes; twigs used to treat toothache, and leaves to cure sores and ulcers.

Jatropha podagrica

Guatemala rhubarb; tartogo; Australian bottle plant

Family *Euphorbiaceae* **Place of origin** Colombia, Central America
Description A curious erect, branched shrub to about 45 cm (1½ ft) with swollen, gouty bases to each branch. **Leaves:** alternate, long stemmed, rounded, with 3 to 5 lobes, about 30 cm (1 ft) across. **Flowers:** small, in cymes on long, smooth, terminal stalks, orange red.
Season Spring, summer.

Lagerstroemia indica

Crape myrtle

Family *Lythraceae* **Place of origin** China
Description A widely cultivated shrub or small tree to 6 m (20 ft) especially in the southern United States and the Caribbean. Bark smooth, grey with patches of green and cream, sometimes pinkish. **Leaves:** opposite, alternate, or in whorls of 3, entire, elliptic to oblong, up to 5 cm (2 in) long. **Flowers:** showy, in sprays, each about 5 cm (2 in) across, delicate looking with 5 to 7 fringed, crumpled petals with many stamens, usually rosy-pink, but occasionally white or purplish.
Season Summer.
Remarks Another species, *L. speciosa,* called pride of India, has similar flowers on 45 cm (1½ ft) panicles and grows to 18 m (60 ft). It has extremely tough timber resistant to salt water, so is much used for boats, wharf posts and the like, also furniture.

Ixora fulgens

Jatropha podagrica

Lagerstroemia indica

Lantana camara *(L. aculeata)* ♄⌀◇
Shrub verbena; yellow sage; lantana

Family *Verbenaceae* **Place of origin** Tropical America
Description A troublesome weed in some countries, particularly
Kenya, India, South Africa and parts of Australia, but a popular
bedding plant in the subtropics or pot plant in cool climates. A
rambling, prickly stemmed, hairy shrub, evergreen, from 1·2–3 m
(4–10 ft) high. **Leaves:** opposite or whorled, simple, oval or oval-
oblong with pointed tips, toothed edges, the many veins giving them
a wrinkled appearance. The leaves when bruised smell like black
currants (*Ribes nigrum*). **Flowers:** axillary, stalked, grouped in
round heads about 2·5–5 cm (1–2 in) across, variously coloured, for
example pink or yellow changing with age to red or orange.
Season Summer.
Remarks The leaves have been used medicinally as a stimulant or
tonic.

Leptospermum scoparium *(L. floribundum)* ♄⌀✿
New Zealand tea tree; manuka

Family *Myrtaceae* **Place of origin** New Zealand, Tasmania
Description The commonest shrub of New Zealand where it adapts
to a wide range of habitats, from dry land to swamps. Height
variable but can grow to 6 m (20 ft) in a good situation. Leaves, young
shoots, flowers and seeds all aromatic. **Leaves:** alternate, evergreen,
stiff, downy, narrowly lanceolate to ovate, 12 mm ($\frac{1}{2}$ in) long.
Flowers: round, five-petalled, white or pink, 12 mm ($\frac{1}{2}$ in) across,
solitary in leaf axils. Cultivars are normally used in gardens; forms
with larger, variously coloured or even double flowers. The rich
red double 'Ruby Glow' is one of the best of these.
Season Spring and summer.
Remarks Often used as a hedging plant, and wood made into tool
handles, fencing posts and windbreaks.

Leucadendron discolor ♄⌀◇
Sunshine bush; gold-tips

Family *Proteaceae* **Place of origin** South Africa
Description An evergreen shrub growing to 1·8–2·1 m (6–7 ft).
Leaves: sessile, stem-clasping, entire, leathery, about 5 cm (2 in)
long and 12 mm ($\frac{1}{2}$ in) wide, oval-oblong, often rimmed with red.
Flowers: male and female distinct, both cone-like, male most con-
spicuous, the woody bracts golden-yellow at their base and red
above. Female cones are green, tinged with red and yellow.
Season Spring.

Lantana camara

Leptospermum scoparium

Leucadendron discolor

Leucospermum conocarpodendron ΨØ◇
(*L. conocarpum*)

Pincushion flower; cripple-wood; kreupelhout

Family *Proteaceae* **Place of origin** South Africa
Description Beautiful evergreen shrub, in flower for months. Height 3 m (10 ft) with a spread of 1·2–1·5 m (4–5 ft). **Bark**: rough. **Leaves**: crowded, entire, leathery, sessile, ovate, to 9 cm (3½ in) long, half as wide, grey-green. **Flowers**: showy, in large round heads about 10 cm (4 in) across, clustered at centre to a dome, individual blooms varying in colour from apricot to flame or salmon-pink. They have prominent, curled styles which make the inflorescence resemble a waxy pincushion.
Season Spring, summer, intermittently all through the year.
Remarks Flowers long-lasting in water. Bark used for tanning.

Malvaviscus arboreus (*M.a.* var. *mexicanus; M. mollis; M. penduliflorus*) Ψ ⚘ ⊸

Sleepy mallow; wax mallow; giant fire dart

Family *Malvaceae* **Place of origin** Mexico
Description Commonly cultivated shrub in the West Indies and southern United States, either alone or as a hedging plant. Much branched and variable from 1–3 m (3–10 ft) high. **Leaves**: alternate, simple, lanceolate to ovate, three-lobed, toothed edges, sometimes unlobed, 9 cm (3½ in) long by 6 cm (2½ in) wide. **Flowers**: solitary or several together in leaf axils, brilliant red, funnel-shaped but narrow at mouth like a loosely furled umbrella, pendulous, 2·5–5 cm (1–2 in) long.
Season Summer, autumn.

Mussaenda erythrophylla Ψ Ø ✳

Family *Rubiaceae* **Place of origin** Tropical West Africa
Description An erect or sometimes climbing shrub to 9 m (30 ft). **Leaves**: opposite, stalked, roundish ovate to elliptic, about 15 cm (6 in) long, slightly hairy beneath. **Flowers**: in groups of 3 or 4, small, pale yellow, set off by one leaf-like bright scarlet sepal over 8 cm (3 in) long and nearly as much across.
Season Spring and summer.
Remarks In India the wood of *M. glabrata,* a similar species, but with white bracts, is made into spoons and other items, designed to avert the evil eye. Milch cattle are tied to pegs of the wood for the same reason. Chewing the roots is reputed to stimulate the appetite.

Malvaviscus arboreus

Mussaenda erythrophylla

Leucospermum conocarpodendron

Mussaenda philippica 'Donna Aurora' Ψ⌀✳

Donna Aurora; flag bush

Family *Rubiaceae*   **Place of origin** Philippines
Description A small evergreen tree of around 5·4 m (18 ft). **Leaves:**
opposite, roundish, ovate to elliptic, about 15 cm (6 in) long. **Flowers:**
the type species has one large white bract and clusters of small
yellow flowers, but 'Donna Aurora' (which was developed in the
Philippines) is double, producing round clusters of ivory bracts
about 20 cm (8 in) diameter at the branch ends. These are very
prolific and weight the bushes like bunches of small handkerchiefs.
Constant cutting back induces more flowers.
Season Almost continuous in the tropics.

Nerium oleander (*N. odorum; N. indicum*) Ψ⌀ or ✳◁▭

Oleander; rosebay

Family *Apocynaceae*   **Place of origin** Mediterranean region east
to Japan
Description Evergreen shrub, 1·8–6 m (6–20 ft) high with long,
slender, upright branches. **Leaves:** in pairs or whorls of 3, leathery,
grey-green, oblong-lanceolate, 20–25 cm (8–10 in) long. **Flowers:** in
terminal bunches each 2·5–5 cm (1–2 in) across, funnel-shaped with
5 lobes, fragrant, various colours from pink to red, white, peach,
yellow, also double.
Season Summer.
Remarks All parts of the plant are poisonous if eaten, yet the flowers
produce good honey. Concoction of roots used to treat ringworm.

Ochna serrulata (*O. japonica; O. multiflora*) Ψ⌀✳

Bird's eye bush; mickey mouse plant; carnival bush

Family *Ochnaceae*   **Place of origin** South Africa
Description An evergreen shrub of 90–150 cm (3–5 ft), colourful
over a long period with its combination of flowers, bright fruits,
dark green adult and reddish young foliage. **Leaves:** alternate,
simple, leathery, narrowly elliptic with serrated edges, 6 cm (2½ in)
long. **Flowers:** several together on short spurs, yellow, buttercup-
like, with 5 to 6 sepals. **Fruits:** round, one-seeded, green, then
scarlet and finally black, persisting for long periods.
Season Spring and summer, irregularly at other seasons.
Remarks The roots of some species are applied as an antidote to
snake bites, and poultices from the leaves used to heal wounds.

Mussaenda philippica 'Donna Aurora'
Nerium oleander

Ochna serrulata

Opuntia vulgaris (*O. monacantha*) Ψ ⊗

Prickly pear; Barbary fig; Irish mittens; tuna

Family *Cactaceae* **Place of origin** Brazil, Argentine
Description Cactus, often used to make stockproof hedges. **Stems:**
jointed, with flat, round or oblong pads about 30 cm (1 ft) long,
narrower at base and joined to others. Grows up to 6 m (20 ft).
Leaves: small and rarely seen, spines on pads which should be
handled cautiously as, being barbed, can hook in the skin. **Flowers:**
yellow or reddish, fluffy centred, 8 cm (3 in) across. **Fruits:** green
turning red, pear-shaped. Another common opuntia is *O. ficus-
indica*, the spineless Indian fig, with 8–10 cm (3–4 in) yellow flowers
and edible red or purple fruits.
Season Summer.
Remarks The prickly pear is a host for the cochineal insect from
which a red dye is extracted.

Pandanus tectorius Ψ ⌀ ◇

Screw pine

Family *Pandanaceae* **Place of origin** Old world tropics
Description Shrubs mostly of coastal regions, frequently branched,
on aerial, stilt-like stems which root into the ground, with tufts of
leaves on top. Some may reach 18 m (60 ft), but most are under 6 m
(20 ft). **Leaves:** long and narrow, often 1·2–1·5 m (4–5 ft) long,
parallel veined, usually twisted in spirals on the stems, hence the
name screw pine. **Flowers:** in large heads, in spathes, male white,
females developing to large, woody, green, pineapple-like fruits.
Season Erratic, mostly summer.
Remarks Leaves used for making bags, sacks, mats and thatching.

Pedilanthus tithymaloides Ψ ⌀ ◇

Redbird flower; devil's backbone; slipper plant; Jewbush; Jacob's
ladder

Family *Euphorbiaceae* **Place of origin** Mexico, South America
Description Succulent, tropical shrub growing to 1·8 m (6 ft) with
fleshy, cylindrical, zig-zag stems containing milky sap. **Leaves:**
alternate, ovate, white-edged, to 10 cm (4 in) long. **Flowers:** in dense
terminal inflorescences, individually like tiny red birds, spurred,
lop-sided, about 2 cm ($\frac{3}{4}$ in) long.
Season Flowers mainly in summer, but intermittently throughout
the year.
Remarks There are two cultivars—'Nana Compacta' with dark
green leaves, and 'Variegatus' with green, white and red leaves.

Opuntia vulgaris
Pedilanthus tithymaloides

Pandanus tectorius

Photinia serrulata   Ψ⌀✳&🜨

Photinia

Family *Rosaceae*   **Place of origin** Japan
Description Evergreen shrub or tree, according to method of pruning. Left unpruned will reach 6–9 m (20–30 ft). **Leaves:** alternate, simple, oblong or elliptic, 5–8 cm (2–3 in) long, dark green, leathery, young leaves brilliant red. **Flowers:** carried in large, flat or rounded clusters, 10–15 cm (4–6 in) across, white, five-petalled.
Season Late spring.
Remarks Makes an attractive hedge, often seen in Australia.

Plumbago auriculata *(P. capensis)*   Ψ⌀✳&🜨

Leadwort; plumbago

Family *Plumbaginaceae*   **Place of origin** South Africa
Description A robust shrub, up to 9 m (30 ft), evergreen in the tropics; deciduous in cooler climates where it is sometimes grown in greenhouses. Woody, branching stems. **Leaves:** alternate, entire, oblong or spoon-shaped, smooth, light green with a tapering point, about 4 cm (1½ in) long. **Flowers:** in large rounded trusses at tips of branches, phlox-like, sky-blue with 5 spreading petals, individually about 2 cm (¾ in) across, calyx sticky.
Season Late summer and autumn.
Remarks Used for hedging purposes in Africa. Forms with white or pink flowers exist but are normally less robust.

Protea barbigera   Ψ⌀◇

Queen protea; woolly-bearded protea

Family *Proteaceae*   **Place of origin** South Africa
Description A popular and showy protea, being increasingly cultivated. Evergreen shrub of spreading habit, up to 135 cm (4½ ft). **Leaves:** alternate, entire, oval, leathery. **Flowers:** distinctive from most proteas so easily recognized, varying in colour from lemon to pink or red, but always with a very dark, almost black, woolly centre. Outer bracts edged with down like a powder-puff. Blooms up to 20 cm (8 in) across.
Season Spring.

—

Photinia serrulata

Plumbago auriculata

Protea barbigera

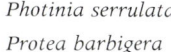

Protea cynaroides ♆ ⬭ ◇
Giant protea

Family *Proteaceae*   **Place of origin** South Africa
Description Branching shrub, now being increasingly planted in the subtropics for its spectacular flowers, which are often cut for decoration or dried in floral arrangements. Leafy branches spring directly from the ground, up to 1·5–1·8 m (5–6 ft) high. **Leaves:** alternate, entire, leathery, oval, to 13 cm (5 in) long, smooth on long stalks. **Flowers:** terminal, large and showy, 15–20 cm (6–8 in) across, normally shell-pink but varying between white to deep red, cup-shaped, with a coating of silky hairs, full of perianth segments and stamens forming a pinkish cone.
Season Spring.

Punica granatum ♆ ⬭ ⬯
Pomegranate; granada

Family *Punicaceae*
Place of origin South East Europe to Himalayas
Description Shrub or small tree widely cultivated for its fruit. Can grow 4·5–6 m (15–20 ft) with spiny branches. **Leaves:** opposite, entire, oblong, 2·5–9 cm (1–3½ in) long, polished upper surface. **Flowers:** bright scarlet with crinkled petals, tubular behind, about 2·5 cm (1 in) across. **Fruits:** large, 8-10 cm (3–4 in) across, thick-skinned, deep yellow flushed red, full of seeds embedded in juicy, edible pulp. Double white and double red forms are occasionally seen in gardens.
Season Summer to early autumn.
Remarks Juice from fruits used to make a drink called Grenadine; bark used medicinally.

Ricinus communis ❋ ❋
Castor oil plant; castor bean; higuerilla

Family *Euphorbiaceae*   **Place of origin** Tropical Africa
Description An annual widely naturalized all over the tropics; small specimens sometimes planted as 'dot' plants in temperate regions. Height normally around 1·5 m (5 ft) but in the tropics can become a small tree. **Leaves:** alternate, palmate with 5 to 12 lobes, long-stemmed but variable in size and colour from green to crimson-brown. **Flowers:** grouped in terminal panicles, green, no petals but many stamens on male flowers, female flowers distinct and borne towards apex of inflorescence. **Fruits:** poisonous, each about 2·5 cm (1 in) long, smooth-skinned or prickly, full of seeds.
Season Summer.
Remarks Oil extracted from seeds used medicinally, as a lubricant, and in paints.

Protea cynaroides

Punica granatum

Ricinus communis

Russelia equisetiformis (R. juncea) Ψ ⌀ ✳ &⊏

Fountain plant; fountain bush; coral plant

Family *Scrophulariaceae*   **Place of origin** Mexico
Description An interesting small shrub commonly planted in
raised situations so that its smooth, drooping, green, rush-like
stems with their myriads of flowers can be seen to advantage.
Height no more than 1·2 m (4 ft). **Leaves:** in whorls of 3 to 6, very
small, elliptic. **Flowers:** plentiful, in clusters of 1 to 3, narrowly
tubular with 5 segments, about 2·5 cm (1 in) long, brilliant scarlet.
Season From spring onwards, usually some flowers in evidence.

Solanum mammosum Ψ ⌀ 器

Nipplefruit

Family *Solanaceae*   **Place of origin** Tropical America
Description Plant curiosity grown mainly for its ornamental fruits
which can be dried. Short, hairy shrub, 60–135 cm (2–4½ ft) high,
branching. Short life-span so usually grown as an annual. **Leaves:**
palmate, 14 cm (5½ in) long and as much across, spiny along the main
and lateral veins. **Flowers:** solitary or few, in leaf axils, small and
purple with reflexed petals round a yellow, central staminal column
—something like a potato flower, about 4 cm (1½ in) across. **Fruits:**
large, 8 cm (3 in) long, 5 cm (2 in) wide, yellow or orange, shaped like
an inverted pear with five nipple-like appendages near the stem.
Season Summer.
Remarks Leaves reputedly used in South America to treat kidney
and bladder ailments. Fruits poisonous to eat.

Tecomaria capensis Ψ ‡ ⊲⊃

Cape honeysuckle

Family *Bignoniaceae*   **Place of origin** South Africa
Description A smooth, rather rambling, evergreen shrub of 1·8–
2·7 m (6–9 ft). **Leaves:** opposite or occasionally in threes, pinnate and
15 cm (6 in) long with 5 to 9 5 cm (2 in) leaflets which may be elliptic,
rounded or oval and toothed at the edges. **Flowers:** massed in
groups at the ends of the upstanding branches, individually narrow,
funnel-shaped, with 5 turned-back petals, fiery orange-red—or
light yellow in some plants.
Season Summer.
Remarks Grown as a hedging plant in southern United States.

Russelia equisetiformis

Solanum mammosum　　　　*Tecomaria capensis*

Telopea speciosissima

Waratah

Family *Proteaceae*　**Place of origin** New South Wales, Australia
Description Upright evergreen shrub to 2·4 m (8 ft), with a number of stems coming from a woody base. **Leaves:** alternate, entire, obovate with toothed edges, 20–25 cm (8–10 in) long, tough and leathery. **Flowers:** spectacular, crimson-scarlet, about 10 cm (4 in) across, in dense, cone-like heads with an involucre or collar of bright scarlet bracts. **Fruit:** a leathery follicle about 10 cm (4 in) long, full of winged seeds.
Season Spring.
Remarks The waratah is the national flower of Australia and strictly protected in the wild.

Tibouchina urvilleana

(*T. semidecandra; Lasiandra macrantha*)

Glory bush; princess flower

Family *Melastomaceae*　**Place of origin** Brazil
Description Handsome evergreen shrubs growing 4·5–6 m (15–20 ft) high, but flowering when quite small. Occasionally seen in greenhouses in temperate countries. **Leaves:** opposite, entire oblong-ovate, 8–15 cm (3–6 in) long, 4 cm (1½ in) wide, velvety to the touch, downy beneath, main veins running in parallel fashion the length of the leaf. **Flowers:** brilliant purple with a velvety sheen, large, 8 cm (3 in) across, five-lobed, long protruding anthers, also purple. Blooms borne in terminal sprays.
Season Summer.

Yucca aloifolia 'Marginata'

Spanish bayonet; dagger plant

Family *Agavaceae*　**Place of origin** Mexico, West Indies, southern United States
Description An erect, evergreen, slender-stemmed shrub of 3–6 m (10–20 ft). **Leaves:** to 60 cm (2 ft) long, 6 cm (2½ in) wide, strap-shaped with very sharp points. **Flowers:** borne on 60–90 cm (2–3 ft) panicles, creamy-white, often purple tinged at base, around 5 cm (2 in) across. 'Marginata' has yellow margins to the leaves.
Season Attractive all year round.

Telopea speciosissima *Tibouchina urvilleana*

Yucca aloifolia 'Marginata'

3
Climbers/Vines

Allamanda cathartica   Ψ ⌀ ⊲

Golden trumpet; golden ball; flor de Manteiga

Family *Apocynaceae*   **Place of origin** Tropical South America
Description Widely grown, vigorous climber to 9m (30ft), usually
seen on walls and fences or sometimes trained as a small shrub.
Leaves: opposite or in whorls of 3 or 4, oval-oblong, 10–15cm
(4–6in) long, glossy. **Flowers:** funnel-shaped, narrowing to a
slender tube, five-parted, golden yellow.
Season Summer to autumn profusely; intermittently all year.
Remarks Stems contain a milky sap which can irritate eyes or lips
if carelessly handled. Sometimes used as an emetic and a purgative.

Antigonon leptopus   Ψ⌀ ✳

Coral creeper; love vine; Honolulu vine; confederate vine; coralita;
chain of love; queen's jewels; queen's wreath; mountain rose

Family *Polygonaceae*   **Place of origin** Mexico, central America
Description Widespread in the tropics. A pretty, tuberous-rooted,
deciduous creeper ascending by means of tendrils up to 12m (40ft),
or, in the absence of supports, scrambling across the ground. **Leaves:**
alternate, entire, arrow-shaped or heart-shaped with pointed tip,
about 8cm (3in) long, bright green, slightly wavy with furrowed
surface. **Flowers:** small and globular in tangled masses on long
racemes which terminate in tendrils that have 3 hooks, bright pink.
There is a white form 'Alba'.
Season Autumn, winter.
Remarks The tubers are edible.

Allamanda cathartica
Antigonon leptopus

Beaumontia grandiflora *(Echites grandiflora)* Ψ🍃◁

Nepal trumpet flower; herald's trumpet; Easter lily vine

Family *Apocynaceae* **Place of origin** India
Description A woody climbing shrub, normally to about 3 m
(10 ft) or so but can grow to 15 m (50 ft) with rusty haired branches.
Leaves: opposite, entire, to 20 cm (8 in) long, leathery, egg-shaped
to oval-oblong with prominent whitish veins. **Flowers**: fragrant,
large, to 10 cm (4 in) long and 10 cm (4 in) across, lily-like or trumpet-
shaped, in clusters of 3 to 9, each opening at a different time, milk
white often tipped with red, veins greenish; buds reddish brown.
Season Spring.

Bougainvillea glabra Ψ🍃✳

Paper flower

Family *Nyctaginaceae* **Place of origin** Brazil
Description A vigorous spiny climber, widely planted, clambers
up tall trees and buildings but may also be trained as standards or
made into hedges. **Leaves**: alternate, entire, bright green, deciduous, roundish ovate. **Flowers**: small white, sitting in centre of 3
large, brilliant magenta bracts, grouped in axillary and terminal
panicles, 15–23 cm (6–9 in) long.
Season Late summer and early autumn but most of the year in some
areas.
Remarks The many cultivars and hybrids are grouped as *B ×
buttiana*. These have orange, lemon, pink or white bracts, some are
double and there is a variegated leaved form.

Clianthus puniceus Ψ‡🍃

Parrot's bill; parrot's beak; kaka beak

Family *Leguminosae* **Subfamily** *Faboideae*
Place of origin New Zeland
Description An evergreen climber up to 3·6–4·5 m (12–15 ft),
grown for its spectacular flowers. **Leaves**: alternate, pinnate,
8–15 cm (3–6 in) long with 12 to 24 opposite, oblong leaflets, 1·2–
2·5 cm (1½–1 in) long. **Flowers**: in pendent trusses, individually
canoe-shaped, bright red, about 8 cm (3 in) long. There is a white
variety.
Season Summer.
Remarks Practically extinct in the wild, the species was saved by
Maoris who cultivated it long before the coming of the white man.

Beaumontia grandiflora

Bougainvillea glabra

Clianthus puniceus

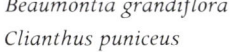

Clytostoma callistegioides

(*Bignonia speciosa; B. callistegioides*)
Argentine trumpet vine; mauve bignonia; love charm

Family *Bignoniaceae*   **Place of origin** Brazil, Argentine
Description An evergreen climber up to 3 m (10 ft) or more with simple tendrils. **Leaves:** opposite, with a pair of entire leaflets, about 8 cm (3 in) long, elliptic oblong, wavy-edged, shiny, bronzed when young. **Flowers:** funnel-shaped widening to campanulate, 6 cm (2½ in) long and wide, five-lobed, mauve or lavender streaked with violet or purple, creamy at base, clustered or in pairs in axils of leaves. Foliage may drop in cold winters.
Season Early summer.

Cobaea scandens

Cup and saucer plant; Mexican ivy; monastery bells; cathedral bell

Family *Polemoniaceae*   **Place of origin** Mexico
Description A vigorous perennial climber often grown as a half-hardy annual in cool climates. In the tropics will grow 6–9 m (20–30 ft) over trees and other supports. **Leaves:** alternate, pinnate with 4 to 6 leaflets, each up to 10 cm (4 in) long, elliptic or oblong. **Flowers:** on long stems, bell-shaped, 4 cm (1½ in) across, greenish at first developing to violet-purple, set off by green, saucer-like calyces. 'Alba' has greenish white flowers and there is another with variegated foliage.
Season Late summer and autumn.

Gloriosa rothschildiana

Climbing lily; glory lily

Family *Liliaceae*   **Place of origin** Uganda, Kenya
Description Climbing members of the lily family, ascending by means of tendrils at the tips of the leaves. Herbaceous perennials with tuberous rootstocks and weak, 90–120 cm (3–4 ft) stems. **Leaves:** long and tapering, alternate, opposite or sometimes whorled, ovate-lanceolate, about 13 cm (5 in) long. **Flowers:** showy, on long stems, yellow and crimson with yellow bases and 6 long, narrow perianth segments, wavy margined, 8 cm (3 in) long.
Season Mid-summer.

Clytostoma callistegioides   *Cobaea scandens*
Gloriosa rothschildiana

Hylocereus undatus (*Cereus undatus*) Ψ 𝔐

Night-blooming cereus; queen of the night; Honolulu queen

Family *Cactaceae* **Place of origin** West Indies
Description An exquisite, night-blooming, epiphytic cactus with green, snake-like, triangular, climbing stems with hairy margins and small spines. Widespread through the tropics where it often grows on palmettos and various trees. **Flowers:** large, 30–35 cm (12–14 in) long, white with many yellow stamens and rosy brown sepals, richly scented. **Fruits:** red, spineless, 10–13 cm (4–5 in) across, edible.
Season Summer.
Remarks Enchanting on a moonlit night.

Ipomoea acuminata (*I. learii*) Ψ ⌀ or ⌘ ⊲

Morning glory; dawn flower

Family *Convolvulaceae* **Place of origin** Tropical America; widely naturalized in tropics
Description Well-known, vigorous, smooth-leaved, perennial climber, to about 3 m (10 ft). **Leaves:** alternate, entire or three-lobed, broadly heart-shaped or rounded, to 13 cm (5 in) long, downy beneath. **Flowers:** large, funnel-shaped, about 13 cm (5 in) long, fine rich blue, opening in the morning to fade by mid to late afternoon. White and pink forms are known.
Season Summer through to autumn.
Remarks Closely allied to and often confused with *I. tricolor* (*I. rubro-caerulea; I. violacea*), china blue or purplish, the parent of many cultivars like 'Heavenly Blue' and the white 'Pearly Gates'. The seeds of some species have hallucinogenic properties, used in religious ceremonies by the Aztecs.

Lapageria rosea Ψ ⌀ ⊲

Chilean bellflower; copihue

Family *Philesiaceae* **Place of origin** Chile
Description A handsome evergreen climber, to a height of 3–4·5 m (10–15 ft). **Leaves:** alternate, simple, dark shiny green, tough and leathery, ovate-lanceolate, 5–8 cm (2–3 in) in length, conspicuously veined. **Flowers:** large and showy, pendulous, waxy, oblong bell-shaped with 6 segments, rosy-crimson spotted rose, 8 cm (3 in) long, by 5 cm (2 in) wide. There are also white and crimson forms.
Season Summer.
Remarks This is the national flower of Chile.

Hylocereus undatus

Ipomoea acuminata

Lapageria rosea

Mandevilla splendens *(Dipladenia splendens)* Ψ⌀⊲

Pink allamanda

Family *Apocynaceae* **Place of origin** Brazil
Description Handsome, woody climber from 3–4·5 m (10–15 ft), with finely hairy stems and milky sap. **Leaves:** opposite, broadly elliptic, to 20 cm (8 in) long, 5–10 cm (2–4 in) wide, stalkless or practically so. **Flowers:** in axillary racemes of 3 to 5, large 8–10 cm (3–4 in) across, rich rose pink with a darker throat, developing with age to rosy madder, funnel-shaped with 5 petal lobes.
Season Spring and summer.

Mutisia spinosa Ψ⌀✺

Family *Compositae* **Place of origin** South America
Description Mutisias are climbing shrubs with large, showy, daisy-like flowers. *M. spinosa (M. retusa)* grows to 6 m (20 ft). **Leaves:** alternate, simple, evergreen, oblong to 5·5 cm ($2\frac{1}{4}$ in), ending in a tendril, stalkless. **Flowers:** pink with yellow centres, 6 cm ($2\frac{1}{2}$ in) across. *M. decurrens* is similar but has brilliant orange flowers 10–13 cm (4–5 in) across; *M. oligodon,* satiny-pink has 45 cm ($1\frac{1}{2}$ ft) stems, and *M. clematis* growing to 6 m (20 ft) has orange-scarlet flowers 8 cm (3 in) across.
Season Summer.

Passiflora edulis Ψ⬡❀

Passion fruit; passion flower; purple granadilla

Family *Passifloraceae* **Place of origin** Brazil
Description Widely cultivated in tropical and subtropical countries for their fruits, but also used as ornamentals. Deciduous vines climbing by means of tendrils. **Leaves:** alternate, three-lobed, smooth, tapering at points, shiny on top. **Flowers:** round, 5–8 cm (2–3 in) across, white petals with fringes of white and purple filaments inside surrounding central staminal column. **Fruit:** size of a small hen's egg, greenish-yellow to purple.
Season Summer.
Remarks Juice and pulp of fruits used for fruit salads, ice-creams, etc. There are about 500 species of passion flower, mostly South American. The various parts of the flowers are fancifully compared to the disciples, apostles, the Trinity and crucifixion—hence the name passion flower.

Mandevilla splendens

Mutisia spinosa

Passiflora edulis

Passiflora mollissima (*Tacsonia mollissima*) Ψꝶ⊗
Curuba; banana passion fruit

Family *Passifloraceae* **Place of origin** Colombia, Venezuela
Description A handsome climbing plant which in Colombia especially or where naturalized elsewhere clambers over hedges and shrubs or creeps along roadsides. **Stems:** cylindrical, strong, downy, reddish-yellow. **Leaves:** alternate, three-lobed, downy with toothed margins, 5–10 cm (2–4 in) long, green but reddish beneath. **Flowers:** pendent, 6–8 cm (2½–3 in) across, with a long calyx tube with 3 large sepals, white inside; petals reddish-pink. **Fruits:** long, downy, oval, yellow, 6 cm (2½ in) long.
Season Spring and summer.
Remarks Curubas in Colombia are made into a refreshing drink, the fruits being extremely juicy.

Petrea volubilis Ψ⊘✳
Queen's wreath; purple wreath

Family *Verbenaceae* **Place of origin** West Indies, Mexico, central America
Description Beautiful evergreen climber, often seen festooning walls, trellises and fences. Sometimes trained as a bush, 1·2–1·8 m (4–6 ft) tall, but will clamber naturally to 6–9 m (20–30 ft). **Leaves:** opposite, entire, short-stalked, elliptic, 2·5–20 cm (1–8 in) long, up to 8 cm (3 in) wide. **Flowers:** in axillary racemes up to 30 cm (12 in) long on a robust specimen. Individual blooms star-shaped, rich purple with 5 lobes, backed by 5 larger, oblong, lilac-mauve sepals. They then look like semi-double flowers but when the inner blooms fall the sepals remain for long periods. There is also a white flowered form called 'Albiflora'.
Season Spring, early summer, sparsely late summer.

Pyrostegia venusta (*P. ignea; Bignonia venusta*) Ψꝶ✳
Flame vine; flame flower; flaming trumpet; golden shower

Family *Bignoniaceae* **Place of origin** Brazil, Paraguay
Description Vigorous evergreen climber, trained along fences, over tree stumps, old buildings, often ascending to the tops of tall trees. **Stems:** ribbed, green, strong. **Leaves:** in pairs, on long stalks, shining, each leaf having 2 or 3 ornate leaflets, about 6 cm (2½ in) long; tendrils on the centre leaflet are used for climbing. **Flowers:** spectacular, in dense drooping clusters, brilliant orange, tubular with 5 reflexed, white-edged lobes, individually about 8 cm (3 in) long.
Season Spring.

Passiflora mollissima

Petrea volubilis

Pyrostegia venusta

Quamoclit coccinea (*Ipomoea coccinea*)   Ψ⌀ ⊲◖

Red morning glory; star ipomoea

Family *Convolvulaceae*   **Place of origin** North America
Description A slender climber of 2·4–3 m (8–10 ft). **Leaves:** alternate, entire, oval heart-shaped, about 15 cm (6 in) long. **Flowers:** small, bright scarlet, funnel-shaped, yellow throat, about 4 cm (1½ in) long.
Season Summer.

Quamoclit lobata   Ψ� ⊲◖

(*Q. mina; Mina lobata; Ipomoea versicolor*)

Family *Convolvulaceae*   **Place of origin** South Mexico
Description An attractive herbaceous climber, up to 4·5–6 m (15–20 ft). **Leaves:** alternate, palmately three-lobed, rounded at base or nearly entire, about 8 cm (3 in) across. **Flowers:** many, in short spikes clustered together in leaf axils, often one-sided, individually forming a swollen tube 2–2·5 cm (¾–1 in) long with the stamens protruding, crimson fading to orange and then yellow.
Season Summer.

Quisqualis indica   Ψ⌀✳ & ▭

Rangoon creeper

Family *Combretaceae*   **Place of origin** New Guinea
Description A common climber of tropical regions with some eccentricities. It starts life as an upright shrub, then sends out runners which climb. This induced early botanists to call it 'quis qualis', i.e. 'which, what?'. **Leaves:** opposite, oblong, pointed, deeply veined, from 2·5–10 cm (1–4 in) long. **Flowers:** distinctive and easily recognizable, in drooping spikes, five-petalled, forming a star, white changing with age to pink or red with lime-green tubes, several inches in length. **Fruits:** black, about 2·5 cm (1 in) long, edible.
Season Spring, early summer.
Remarks The edible leaves taste something like radishes.

Senecio confusus   Ψ⌀❀

Mexican flame vine; orange flame vine; Mexican daisy; Mexican fire vine

Family *Compositae*   **Place of origin** Mexico
Description A showy, evergreen climber, often seen growing over fences and tree stumps. **Leaves:** alternate, simple, thickish, ovate, to 5 cm (2 in) long, edges toothed. **Flowers:** in loose terminal trusses, orange or orange-red with yellow centres, daisy-like, each about 4 cm (1½ in) across.
Season Spring, early summer, intermittently at other times.

Quamoclit coccinea
Quisqualis indica

Quamoclit lobata
Senecio confusus

Solandra guttata Ψ⌀◊

Chalice vine; cup of gold; trumpet plant

Family *Solanaceae* **Place of origin** Mexico, West Indies
Description A spectacular climber up to 9 m (30 ft). **Leaves:** large,
8–15 cm (3–6 in) long, oval, leathery, shiny. **Flowers:** goblet-shaped,
very large, five-petalled, to 25 cm (10 in) across, soft yellow on
opening, becoming golden with age, purple tracings running down
inner petals. Fragrant with a smell of coconut. **Fruits:** fleshy berries
with two seeds.
Season Spring until autumn, intermittently.

Solanum jasminoides Ψ occ. Ψ⌀❋

Potato vine

Family *Solanaceae* **Place of origin** Brazil
Description A shrubby climber; height around 4·5 m (15 ft).
Leaves: alternate, ovate-lanceolate on wiry stems, 2·5–8 cm (1–3 in)
long, entire, smooth. **Flowers:** in clusters of about 10 blooms,
individually 2·5 cm (1 in) across, five-lobed, star-shaped, white or
mauve tinted.
Season Summer, early autumn.

Solanum wendlandii Ψ⌀ or ❋ ❋ & ❀

Potato vine; giant potato creeper; Costa Rican nightshade; paradise
flower

Family *Solanaceae* **Place of origin** Costa Rica
Description A robust climber with large, heavy trusses of flowers.
Ascending to 6 m (20 ft) or more, stems bearing a few hooked
prickles. **Leaves:** large, usually deciduous, alternate, variable,
usually pinnate, up to 25 cm (10 in) long, but upper leaves may be
simple, heart-shaped or three-lobed, bright green. **Flowers:** in
branched clusters, 15–20 cm (6–8 in) across, each flower 6 cm ($2\frac{1}{2}$ in)
across, deep mauve with a bright yellow centre, five-petalled.
Fruits: oval or round, about 8 cm (3 in) across, but rarely seen.
Season Late summer and autumn.

Streptosolen jamesonii *(Browallia jamesonii)* Ψ⌀❋

Firebush; marmalade bush; yellow heliotrope

Family *Solanaceae* **Place of origin** Colombia, Ecuador
Description Showy climber, normally about 1·8 m (6 ft) high.
Frequently grown as a conservatory plant in cooler climates. Likes
plenty of sun. **Leaves:** alternate, entire, ovate, 3 cm ($1\frac{1}{4}$ in) long.
Flowers: in dense clusters, each about 3 cm ($1\frac{1}{4}$ in) long, tubular and
four- to five-lobed, brilliant orange.
Season Summer.

Solandra guttata
Solanum wendlandii

Solanum jasminoides
Streptosolen jamesonii

Strongylodon macrobotrys
Jade vine

Family *Leguminosae* **Subfamily** *Faboideae*
Place of origin Philippines
Description A woody climber, with strong stems growing 12–15 m (40–50 ft) long and 2·5 cm (1 in) in diameter. **Leaves:** large and pinnate with 3 13 cm (5 in) leaflets. **Flowers:** on long-stemmed racemes 60–90 cm (2–3 ft) long; individual blooms, bluish-green, 8 cm (3 in) long, pea-shaped.
Season Spring, early summer.

Thunbergia alata
Black-eyed Susan

Family *Acanthaceae* **Place of origin** Tropical Africa
Description A twining climber, perennial but usually dying down in winter. Grows to height of 3–3·6 m (10–12 ft) in a season. **Leaves:** opposite, entire, triangular or heart-shaped, 8 cm (3 in) long, soft, bright green, toothed. **Flowers:** trumpet-shaped with a flat face, 4 cm (1½ in) across, orange-yellow with purplish-black centre, or white with a dark centre in some garden forms.
Season Mid-summer to autumn.
Remarks Grown as a annual pot plant in temperate countries.

Thunbergia grandiflora
Blue trumpet vine; skyvine; heavenly blue

Family *Acanthaceae* **Place of origin** North India
Description A vigorous climber which makes a dense curtain of foliage superimposed with strings of vivid blue flowers. **Leaves:** opposite and simple but variable in shape according to age. Normally ovate tapering to a long point when young but becoming almost heart-shaped and 15–20 cm (6–8 in) long at maturity. **Flowers:** in drooping racemes, large, 5–8 cm (2–3 in) across at top, five-lobed, lavender-blue with white base, tubular below.
Season Spring but intermittently at other seasons.

Thunbergia mysorensis
(*Hexacentris mysorensis*)

Family *Acanthaceae* **Place of origin** India
Description A twining, woody climber to about 4·5 m (15 ft). **Leaves:** opposite, entire, elliptic, 10–15 cm (4–6 in) long, toothed. **Flowers:** in racemes, yellow, funnel-shaped with purple tube, about 5 cm (2 in) across.
Season Spring.

Strongylodon macrobotrys
Thunbergia grandiflora

Thunbergia alata
Thunbergia mysorensis

4
Waterside Plants

Carpobrotus edulis *(Mesembryanthemum edule)* Ψ ⌀ ✿
Hottentot fig; sour fig; gouna

Family *Aizoaceae* **Place of origin** South Africa

Description A perennial succulent, rather coarse, with long trailing, fleshy stems to 90 cm (3 ft) or more. **Leaves:** opposite, simple, thick, curved, three-angled, tapering to the tip, 8–10 cm (3–4 in) long, 8 mm ($\frac{1}{3}$ in) wide. **Flowers:** large and showy, daisy-like, opening mid-day, 8 cm (3 in) or more across, yellow with silky petals turning pink with age. Purple kinds also common. Widely naturalized in Europe, Madeira, United States and other parts of the world, particularly on sandy shores where they festoon sea walls, cliffs, etc.

Season All summer.

Remarks Fruits brown, fleshy, like miniature figs, are popular jam fruits in South Africa.

Colocasia esculenta *(C. antiquorum esculenta; Caladium esculentum)* Ψ ⌀ ⚲
Taro; dasheen; elephant's ear; eddo; kalo

Family *Araceae* **Place of origin** East Indies

Description Widely grown in the tropics, particularly in the Pacific Islands as a source of food, in moist localities. It is also used as an ornamental in tropical pools. The species is variable and many forms occur with variously marked leaves and stems. Tuberous-rooted. **Leaves:** arise from base, large and shaped like the ears of an African elephant, blades thin, velvety, green, 50 cm (20 in) or so long, on green, violet or reddish, 1·05 m ($3\frac{1}{2}$ ft) stems. **Flowers:** rarely seen, but aroid-like, green or reddish-purple or whitish, 30–38 cm (12–15 in) long.

Season Available most seasons.

Remarks Blanched and forced shoots cooked and used as a winter vegetable; tubers reputedly more nutritious than potatoes, eaten in stews and a variety of dishes, but must be boiled to remove bitter calcium oxalate crystals. Also source of Portland arrowroot.

Carpobrotus edulis
Colocasia esculenta

Cyperus papyrus *(Papyrus antiquorum)* Ψ⌀

Papyrus; paper plant; Egyptian paper reed

Family *Cyperaceae* **Place of origin** North and tropical Africa
Description Plants of great antiquity, reputedly the 'rushes' which sheltered the infant Moses. Often planted in pools or wet soil in tropical and subtropical gardens. Perennial with short rhizome and fibrous roots, triangular stems, 2·4–3 m (8–10 ft) or even higher in places like Uganda. **Leaves:** small, grassy, stems topped by large, round, mop-head umbels of grassy, greenish brown flowers in spikelets. If these fall or are pressed into mud the seeds germinate to make new plants.
Season Summer.
Remarks Papyrus, the world's earliest writing paper was made from the pithy tissues of this plant's flowering stems. These were cut into thin strips and pressed together while still wet.

Eichhornia crassipes *(E. speciosa)* Ψ⌀◇

Water hyacinth

Family *Pontederiaceae* **Place of origin** Tropical America
Description A floating plant, often in dense masses in streams, slow-moving rivers or flat stretches of water. **Leaves:** arranged in rosettes, individually smooth, heart-shaped, about 5 cm (2 in) across, each leaf stalk swollen to a sausage shape and full of spongy tissue and air. These make the plants buoyant. **Flowers:** bright and showy, in spikes, soft lavender blue with gold and deep blue 'peacock' markings on top petal, funnel-shaped, each about 4 cm (1½ in) long. Long, black trailing roots which fish use for spawning purposes. Plant increases by means of runners.
Season Summer.
Remarks An obnoxious weed in some countries, particularly Egypt, Malaya and parts of the United States. Plants fed to cattle in South America.

Hydrocleys nymphoides *(Limnocharis humboldtii;*

H. commersonii; Hydrocleis nymphaeoides) Ψ⌀⊗

Water poppy

Family *Butomaceae* **Place of origin** South America
Description A water plant with long, trailing stems which root into the mud under shallow conditions: **Leaves:** bunched together, several from the same point, floating, simple, fleshy, thick, oval to heart-shaped, 5 cm (2 in) across with longitudinal veins and long-stemmed. **Flowers:** long-stemmed, in clusters, showy but short-lived, bright yellow, 3 petals, 5–6 cm (2–2½ in) across, standing just above the water.
Season Throughout the summer.

Cyperus papyrus
Hydrocleys nymphoides

Eichhornia crassipes

Ipomoea pes-caprae Ψ⌀◁

Beach morning glory; seaside morning glory

Family *Convolvulaceae* **Place of origin** World tropics
Description A common seashore plant found on countless beaches from the Galápagos Islands to Australia's barrier reef islets. It creeps along the ground in long branching strands up to 18 m (60 ft) rooting here and there as it goes. **Leaves:** alternate, simple, smooth, fleshy, rounded, something like a goat's footprint, 5–10 cm (2–4 in) across. **Flowers:** about 5 cm (2 in) across, trumpet-shaped, five-lobed, rosy-purple, closing soon after midday.
Season Spring, but in evidence most seasons of the year.

Nelumbo nucifera *(Nelumbium speciosum)* Ψ⌀❀

Sacred lotus

Family *Nelumbonaceae* **Place of origin** Asia
Description Historic and spectacular water plant. Rhizomatic root. **Leaves:** simple, round, with centrally placed stout stem, 30–90 cm (1–3 ft) tall, glaucous with a waxy surface, water-repellant. **Flowers:** showy, large, up to 30 cm (1 ft) across, peony-shaped, rosy-pink with a prominent ovary in the centre resembling the rose of a watering can. White and carmine forms occur, also doubles. **Fruits:** edible almond-sized nuts.
Season Summer, intermittently spring and autumn.
Remarks A plant with religious connotations for Christians (Ecclesiastes, 11 v. 1) and Buddhists, the latter regarding it as sacred.

Nymphaea ampla Ψ⌀❀

Water lily

Family *Nymphaeaceae* **Place of origin** South America; genus world-wide
Description Popular water plants; some night-blooming. The hardy water lily flowers (of temperate countries) usually float on the water, but tropicals bear their blooms well above the surface and many are sweetly scented. The blue-flowered types are prized for garden pools. These are narrow-petalled and have many filaments in the centre, the same colour as the petals. There are pink, white, yellow and purple garden varieties. Night-bloomers have broader petals and are usually white or red. **Leaves:** variable, simple, round or orbicular, plain or crinkly-edged, green or blotched with chocolate, from 8–45 cm (3–18 in) across according to variety. *N. ampla* from tropical America has large leaves, 40 cm (16 in) across, with black spots on the underside and 13 cm (5 in) white flowers.
Season Summer.
Remarks *N. caerulea* is the blue lotus of the Nile.

Ipomoea pes-caprae   *Nelumbo nucifera*
Nymphaea ampla

Portulaca lutea

Sun plant; eleven o'clock; purslane

Family *Portulacaceae* **Place of origin** South America, but widely distributed in tropics

Description *P. grandiflora* is a soft-stemmed annual, trailing or erect, to about 30 cm (1 ft). **Leaves:** alternate, fleshy, sharp-tasting, rounded, about 2·5 cm (1 in) long. **Flowers:** yellow, round, daisy-like, about 2·5 cm (1 in) across. Garden forms with rose, red, white and bicoloured flowers are common in tropical gardens.

Season summer.

Remarks *P. lutea* (illustrated) is a widespread species of the Pacific region, often found on coral or sandy shores.

Rhizophora mangle

Red mangrove; mangle rojo

Family *Rhizophoraceae* **Place of origin** Tropical America, Caribbean, west coast Africa

Description Tropical trees found only in coastal areas or wetlands subject to tidal flooding. They can grow 24–30 m (80–90 ft) high, spreading outwards by means of many stilt-like, arching, aerial roots which grow down into the soil. These tangled appendages trap mud and debris and in time form mud flats. **Leaves:** opposite, deciduous, large, oval to elliptic. **Flowers:** white turning brown, four-petalled, 1·5 cm (½ in) across. **Fruits:** pencil-like, 30 cm (12 in) long, containing one seed. This germinates while still on the tree, then falls and roots into the mud.

Season Summer.

Remarks The wood is very durable, is able to withstand water and is also used as a source of charcoal.

Victoria amazonica (*V. regia*)

Royal water lily; water platter

Family *Nymphaeaceae* **Place of origin** Tropical South America

Description Very large water plant, rooting in mud but leaves and flowers floating. **Leaves:** simple, on long spongy stems, blades 90–120 cm (3–7 ft) across, circular, green but reddish beneath, also spiny beneath. Older leaves develop upturned rims to about 15 cm (6 in) high which show the red undersides. The similar but hardier *V. cruziana* has green margins 15–20 cm (6–8 in) high. **Flowers:** opening towards evening, fragrant, water lily-like, white, changing on successive nights to pink or red, sepals and flower stalks prickly, to 30 cm (1 ft) across. **Fruits:** pods with many seeds.

Season Late spring and summer.

Portulaca lutea
Rhizophora mangle

Victoria amazonica

5
Miscellaneous
(including low growing and creeping plants)

Agapanthus praecox (A. umbellatus)   Ψ⌀❊
Lily of the Nile; African lily

Family *Amaryllidaceae* (or *Alliaceae* by some authorities)
Place of origin South Africa
Description Well-known bulbous perennials with showy flowers and thick, fleshy roots. There is a number of species and many cultivars, but this is the most common. Height 60–75 cm (2–2½ft) but sometimes taller. **Leaves:** basal, soft, long and narrow, sword-shaped, up to 5 cm (2 in) wide, deciduous. **Flowers:** impressive, in large round umbels, containing between 20 and 100, stalked, 5 cm (2 in) flowers, funnel-shaped, blue. Forms exist with white, pale to deep (near navy) blue, double, and with variegated leaves.
Season Summer to early autumn.
Remarks Flowers good for cutting. Tubers have been used medicinally for heart and intestinal disorders.

Agave americana   Ψ⌀❊
Century plant; maguey; American aloe

Family *Agavaceae*   **Place of origin** Mexico
Description An evergreen plant which resembles a large cactus and forms a rosette of leaves close to the ground. **Leaves:** basal, leathery, grey-green, or cream-edged in 'Marginata' up to 1·2 m (4 ft) long, and about 25 cm (10 in) wide. These are rigid but recurved, with sharp spines at their tips and teeth along the margins. **Flowers:** borne on massive 7·5 m (25 ft) stems which are branched, the greenish-yellow flowers, each 9 cm (3½in) length, massed at the ends of the lateral branches. The plants die after blooming, leaving offsets to replace them, but these take many years to flower—hence the name century plant.
Season Leaves attractive all the year, flowers spasmodic.
Remarks Mexicans produce an intoxicating beer (pulque) from the macerated stems and leaf bases. Most agaves have fibrous foliage; *A. sisalana* for example is the source of sisal hemp.

Agapanthus praecox
Agave americana

Agave attenuata   Ψ⌀✳

Family *Agavaceae*   **Place of origin** Mexico
Description A handsome, rosette-forming succulent commonly planted in coastal districts where it thrives in dry, arid soil. Forms a woody stem, extending to 90 cm (3 ft) or so with age, topped by rosettes of thick, succulent, grey-green leaves, 15–38 cm (6–15 in) wide in the centre with pointed tips. Young plants sprout from the base. **Flowers:** greenish-yellow, funnel-shaped, each about 5 cm (2 in) long, crowded on branching, sturdy, 2·4 m (8 ft) stems.
Season Summer.
Remarks Agave leaf-fibre makes strong twine and ropes. The drug mecogenin, which can be converted to cortisone, is produced from some species.

Alpinia purpurata   Ψ⌀◇

Red ginger

Family *Zingiberaceae*   **Place of origin** Pacific Islands
Description A widely cultivated plant in tropical gardens, particularly in moist shade. Of upright habit it is a leafy plant sometimes to 3·6 m (12 ft) but more commonly 1·2–1·8 m (4–6 ft). **Leaves:** sprouting from underground rhizomes, around 75 cm (2½ ft) long by 15 cm (6 in) wide, strap-like, or about 30 cm (12 in) long growing on the upper parts of the flower stems. **Flowers:** in terminal clusters consisting of bright red persistent bracts over 2·5 cm (1 in) long. These enclose small white flowers. The double form, 'Plena', is illustrated.
Season Spring to late autumn.
Remarks The bracts are often strung into leis in Hawaii and the Pacific Islands. In Samoa they are occasionally fashioned into a chain-mail type ornamental dress.

Alpinia zerumbet *(A. speciosa; A. nutans)*   Ψ⌀▷&◇

Shell ginger; shellflower; pink porcelain lily

Family *Zingiberaceae*   **Place of origin** East Asia, China, Japan
Description A perennial herb growing to 2·7–3·6 m (9–12 ft) with fine foliage and showy flowers. **Leaves:** practically stalkless, lanceolate, 60 cm (2 ft) long and 13 cm (5 in) wide. **Flowers:** large on leafy stems, in drooping racemes, each about 4 cm (1½ in) long with lobed lip, shell pink with golden yellow and red stripes and suffusions.
Season Spring but intermittently at other seasons.
Remarks Common plant of tropical gardens often planted in deep shade.

Agave attenuata

Alpinia purpurata 'Plena'

Alpinia zerumbet

Anigozanthos manglesii   Ψ⌀▷

Kangaroo paws

Family *Haemodoraceae*   **Place of origin** South west Australia
Description A genus of perennial herbs with short rhizomes, whose height and colouring varies according to the species, but their characters are as follows. **Leaves:** basal, linear or narrowly strap-shaped, tough, deep green. **Flowers:** borne on one-sided woolly racemes, individually tubular, long and narrow, from 2·5–8 cm (1–3 in) long, with 6 lobes. As the flower ages it splits down one side and the perianth lobes roll back. *A. flavidus* is one of the tallest at 1·5–1·8 m (5–6 ft), yellowish-green flowers, 2·5–4 cm (1–1½ in) long; *A. manglesii* grows to 90 cm (3 ft) with red and green hairy flowers, each 7 cm (2¾ in) long. *A. preissii*, a small-growing species only 50 cm (20 in) tall, has gold and red flowers, each 5–6 cm (2–2½ in).
Season Spring.

Anthurium andreanum   Ψ⌀Q

Painter's palette; flamingo lily

Family *Araceae*   **Place of origin** Colombia
Description A popular evergreen foliage plant, used in cool climates as a greenhouse pot plant. Height 45–60 cm (1½–2 ft). **Leaves:** basal, long-stalked, deep green, heart- to arrow-shaped with 20–25 cm (8–10 in) blade. **Flowers:** arum-like, waxy red, shiny, flat, often puckered, 8–13 cm (3–5 in) long and 8 cm (3 in) wide with a cylindrical spadix in the centre, 6 m (2½ in) long, pale golden.
Season Summer to autumn.
Remarks The species is parent of many fine hybrids with white, pink, deep red or salmon flowers.

Begonia maculata *(B. argyrostigma)*   Ψ⌀✳

Family *Begoniaceae*   **Place of origin** Brazil
Description A fibrous-rooted, succulent perennial of branching habit and 60–90 cm (2–3 ft) stems, which become woody with age. **Leaves:** alternate, about 13–15 cm (5–6 in) long, orbicular with rounded base and wavy margins, green with white spots, crimson beneath. **Flowers:** plentiful, on drooping, axillary sprays, pale pink, about 12 mm (½ in) across. A white-flowered cultivar, called 'Wightii', grows 1·8–2·4 m (6–8 ft) tall.
Season Summer, intermittently at other times.
Remarks There are many begonias to be found in tropical regions, the fibrous kinds usually growing in shade.

Anigozanthos manglesii
Anthurium andreanum

Begonia maculata

Bromeliads   Ψ⌀⌗ m

Air plants

Family *Bromeliaceae*   **Place of origin** South east United States to South America.

Description A large group of about 60 genera and 1400 species of plants; some live on the ground but many perch on rocks or in tree crevices, where their brilliant flowers and squat leaves are reminiscent of perching birds. Individual flowers vary but all have 3 petals and 6 stamens. The leaves of many grow in rosettes which curve together at their base and form a vase-like shape which will hold water. The chief genera include vrieseas, tillandsias, aechmeas, billbergias, cryptanthus, neoregelias, also pineapples (*Ananas comosus*) and Spanish moss (*Tillandsia usneoides*) p. 130.

Caladium bicolor   Ψ⌀ ⓠ

Elephant's ear; angel wings

Family *Araceae*   **Place of origin** Brazil

Description Popular pot plants in cold climates but used as bedding plants in the tropics, especially in damp borders, for their brightly coloured foliage. Tuberous perennial. **Leaves:** entire, about 30 cm (1 ft) long, usually arrow-shaped or triangular, stalks spotted or coloured, blades flat, soft, marked and suffused with various colours — green, buff, crimson, white, rose and bright red, often ruffled at the edges and long-stemmed. **Flowers:** unimportant but arum-like, green or yellowish. The hybrids are often listed as *C.* × *hortulanum*.

Season Spring to autumn.

Canna indica   Ψ⌀✻

Indian shot; atzera

Family *Cannaceae*   **Place of origin** Tropical America

Description Erect, leafy perennial to 90–120 cm (3–4 ft), with branching rhizomes. **Leaves:** simple, entire, broad, with sheathing petioles, up to 45 cm (1½ ft) long, and 15–20 cm (6–8 in) wide, dark green. **Flowers:** bright red or pink, the lip orange spotted red, about 8 cm (3 in) long. There are also many cultivated cannas with much larger flowers, variously coloured in shades or mixtures of scarlet, crimson, pink, yellow or orange. The leaves of some cultivars are crimson instead of green.

Season All summer.

Remarks The tubers of another species, *C. edulis* are eaten by Andean Indians, who also wrap their new-born babies in the leaves.

Bromeliads spp.

Caladium bicolor

Canna indica

Catharanthus roseus (*Vinca rosea; Lochnera rosea*)

Rose periwinkle; Madagascar periwinkle; old maid   Ψ⌀🏵

Family *Apocynaceae*   **Place of origin** Madagascar, widely naturalized in tropics
Description Soft-stemmed perennial of spreading habit, up to 60 cm (2 ft) high widely planted in tropical gardens, also grown as a pot plant in cool climates. Stems leafy. **Leaves:** opposite, entire, smooth, oblong lanceolate, 2·5–5 cm (1–2 in) long. **Flowers:** funnel-shaped with a gradually expanded tube so that the tops appear saucer-shaped, calyx and corolla five-parted, up to 4 cm (1½ in) across, bright rose-pink, crimson or white.
Season Continuous.
Remarks The plant contains various alkaloids and is used medicinally for some forms of cancer.

Chrysothemis pulchella (*Tussaca pulchella*)   ⌀ ⊲

Dozakie

Family *Gesneriaceae*   **Place of origin** Panama, West Indies
Description A tuberous perennial, popular for bedding or as a pot plant. Height around 30 cm (1 ft). **Leaves:** opposite, simple, oblong-lanceolate, 20–25 cm (8–10 in) long, prominently veined, toothed, rough-corrugated. **Flowers:** axillary, several together, individually 2·5–8 cm (1–3 in) long, orange-yellow with red calyx, tubular with 5 petals.
Season Summer.

Clerodendrum paniculatum   Ψ⌀✳

Pagoda flower

Family *Verbenaceae*   **Place of origin** South east Asia
Description An erect shrub of gardens and waste places, to 90 cm (3 ft). **Leaves:** orbicular, five-lobed, 17–38 cm (7–15 in) wide. **Flowers:** orange-red, in large terminal panicles up to 30 cm (1 ft) long; corolla tube 12 mm (½ in) long. **Fruit:** black berries.
Season Summer.

Catharanthus roseus
Chrysothemis pulchella

Clerodendrum paniculatum

Clianthus formosus   Ψ ‡ ⅏

Sturt desert pea; glory pea; desert pea

Family *Leguminosae*   **Subfamily** *Faboideae*
Place of origin Australia
Description An evergreen, sprawling, desert shrub to 60–90 cm
(2–3 ft). **Leaves:** alternate, pinnate, silvery, covered with silky
hairs, leaflets 1·2–2·5 cm ($\frac{1}{2}$–1 in), oval. **Flowers:** striking, pea-
shaped, several in a cluster, each 8 cm (3 in) long, rich red with a
large purple-black blotch at base of top petal (standard). Often grown
in hanging baskets or grafted on a seedling of *Colutea arborescens,* a
near relative.
Season Spring and summer.

Crinum latifolium var. zeylanicum   Ψ ⬭ ⊲

(*C. zeylanicum; Amaryllis ornata*)
Milk and wine lily

Family *Amaryllidaceae*   **Place of origin** Tropical Asia
Description Handsome bulbous plant with large bulbs, up to
20 cm (8 in) in diameter. **Leaves:** numerous, strap-shaped, 90–150 cm
(3–5 ft) long, 5–6 cm (2–2$\frac{1}{2}$ in) wide. **Flowers:** on 60–90 cm (2–3 ft)
stems, fragrant, in umbels of 10 to 20, individually 8–10 cm (3–4 in)
long and 2·5 cm (1 in) wide, white with purplish red markings.
Season Spring and summer.
Remarks Leaf juices used in India to alleviate ear-ache, and the
bulbs, after roasting, are laid on the skin to ease rheumatic pain.

Dichorisandra thyrsiflora   ⬭ ◇

Queen's spiderwort

Family *Commelinaceae*   **Place of origin** Tropical America
Description Herbaceous perennial with ornamental foliage and
attractive flowers. Height about 60 cm (2 ft), stems reddish purple
with green blotches when young, then green later. **Leaves:** closely
arranged and sheathing the stems, alternate, sharp-pointed, lanceolate, about 30 cm (12 in) long and 10–13 cm (4–5 in) wide, purple
beneath, sometimes white flecked. **Flowers:** in close, erect, terminal
racemes three-petalled, deep blue or bluish-purple with white
towards base.
Season Summer.

Clianthus formosus

Crinum latifolium var. *zeylanicum*

Dichorisandra thyrsiflora

Echinocactus grusonii

Barrel cactus; golden ball cactus; mother-in-law's chair

Family *Cactaceae* **Place of origin** Central Mexico
Description A large, round or barrel-shaped cactus, depressed at the top, which will grow to 90 cm (3 ft) across and is studded all over with formidable, sharp, golden spines. These vary in length from 2·5–5 cm (1–2 in). **Flowers:** solitary, yellow, campanulate on a long tube, 5 cm (2 in) long.
Season Interesting at all seasons.

Echium candicans

Pride of Madeira

Family *Boraginaceae* **Place of origin** Madeira, Canary Islands
Description A number of tall-growing echiums are grown as garden plants in various parts of the world. Most of these are native to Madeira and the Canary Islands, including the blue or white, 1·8 m (6 ft) *E. candicans*, and pink, 2·4–3 m (8–10 ft) *E. wildpretii*. Both are biennial, dying after flowering, shrubby and branching with leafy stems. **Leaves:** alternate, lanceolate, 15–20 cm (6–8 in) long, covered with silvery hair, prominently veined. **Flowers:** tightly packed at the top of long 60–90 cm (2–3 ft) spikes, deep blue, almost stemless, tubular, dilated at top, five-petalled.
Season Spring, early summer.

Eupatorium sordidum (*E. ianthinum*)

Boneset; thoroughwort; mist flower

Family *Compositae* **Place of origin** Mexico
Description A shrubby plant about 1·8 m (6 ft) tall, used in mixed flower borders. **Leaves:** large, ovate to oblong-ovate, coarsely toothed, about 10 cm (4 in) long. **Flowers:** fragrant, in loose clusters, violet-mauve, like tiny shaving brushes, each about 1·2 cm ($\frac{1}{2}$ in) across. *E. atrorubens,* also from Mexico, is similar, with reddish-lilac flowers.
Season Autumn and winter.

Echinocactus grusonii

Echium candicans

Eupatorium sordidum

Furcraea foetida (F. gigantea)   Ψ ⌀ ◇

Mauritius hemp; green aloe

Family *Agavaceae*   **Place of origin** North South America
Description A striking plant in leaf and flower. Succulent, forming impressive leaf rosettes and occasionally flower spikes up to 9 m (30 ft) or more. Plant dies after blooming but renews itself from offsets. **Leaves:** sword-shaped, basal, 1·8–2·4 m (6–8 ft) long, 15–20 cm (6–8 in) wide, margined with a few distant prickles. **Flowers:** in panicles on thick stems, milk white, greenish outside, each with 6 segments and 6 stamens, 4 cm (1½ in) long. 'Mediopicta' (*F. watsoniana*) has cream leaf variegations.
Season All year round; flowers late summer.

Geranium maderense   Ψ ✽ ✽

Madeira stork's bill

Family *Geraniaceae*   **Place of origin** Madeira
Description A striking and robust herbaceous perennial, up to 1 m (3 ft) high. **Leaves:** large, palmately divided, each segment deeply cut almost to the base, the strong stems covered with long hairs. **Flowers:** many, in large, showy, branched inflorescences, sometimes more than 1·5 m (5 ft) across, individual blooms 3–4 cm (1¼–1½ in) across, glowing rosy-purple.
Season Spring until autumn.

Hedychium gardneranum   Ψ ⌀ ◇

Kahili ginger; ginger lily

Family *Zingiberaceae*   **Place of origin** India
Description Perennial herbs noted for their showy flowers and fragrance. Rhizomatous with strong leafy stems, 90–180 cm (3–6 ft) tall. **Leaves:** flat, lanceolate, 30–38 cm (12–15 in) long, to 13 cm (5 in) wide, resembling those of a canna plant. **Flowers:** in rounded spikes at tops of the flower stems, packed with 5 cm (2 in), lemon yellow, long-tubed, two-lobed flowers with bright red, gold-topped very long stamens.
Season Summer.

Geranium maderense

Hedychium gardneranum

Furcraea foetida 'Mediopicta'

Heliconia rostrata

Lobster claw

Family *Heliconiaceae*   **Place of origin** Argentina to Peru
Description Perennial herbs closely related to the banana and usually found in damp localities growing to 1·8 m (6 ft). **Leaves:** large, oblong-acute, 60–120 cm (2–4 ft) long, 25 cm (10 in) wide. **Flowers:** on erect 1·2 m (4 ft) stems, bracts brilliant scarlet and yellow, enclosing small white flowers.
Season Spring; intermittently at other seasons.

Hippeastrum reginae *(Amaryllis reginae)*

Mexican lily; amaryllis

Family *Amaryllidaceae*   **Place of origin** Brazil, tropical America
Description A bulbous plant to 60 cm (2 ft). **Leaves:** basal, strap-shaped, bright green, about 4 cm (1½ in) wide. **Flowers:** on smooth, naked, 30 cm (1 ft) stems, grouped, several together, individually red with greenish-white central markings, 10–13 cm (4–5 in) long.
Season Spring or early summer.

Kalanchoe pinnata *(Bryophyllum pinnatum)*

Air plant; life plant; floppers; curtain plant

Family *Crassulaceae*   **Place of origin** World tropics
Description A smooth, succulent-leaved, hollow-stemmed herb growing to 90–120 cm (3–4 ft). **Leaves:** opposite or in whorls of three but variable, being pinnate with 3 to 5 leaflets at top of stems but lower down are simple, oblong-lanceolate with toothed edges; variable in colour from green to russet, purple blotched beneath, 5–20 cm (2–8 in) long. **Flowers:** pendent, in bunches, tubular and inflated, green and pink, 2·5–4 cm (1–1½ in) long, becoming papery with age.
Season Early summer onwards.

Lampranthus zeyheri

Common purple lampranthus

Family *Aizoaceae*   **Place of origin** South Africa
Description Perennial succulent producing so many flowers that they completely mask the foliage. The blooms need full sun so are generally found in hot, dry gardens, clambering over banks and the like. Height about 60 cm (2 ft), plant bushy. **Leaves:** opposite, cylindrical to triangular, slightly curved, prolific. **Flowers:** daisy-like in appearance, 6 cm (2½ in) across, gleaming red-purple with a single row of outer florets and yellow centre.
Season Spring to summer.

Heliconia rostrata

Hippeastrum reginae

Kalanchoe pinnata

Lampranthus zeyheri

Lotus berthelotii

Parrot's beak; pelican's beak; coral gem; winged pea

Family *Leguminosae* **Subfamily** *Faboideae*
Place of origin Canary Islands
Description Scrambling perennial with a woody rootstock which drapes walls, rocks and bare ground with carpets of silvery foliage spangled with myriads of flowers. **Leaves:** alternate, pinnate about 1·2 cm (½ in) long, with 3 to 7 very narrow, silky hairy leaflets. **Flowers:** brilliant scarlet, pea-like, about 2·5 cm (1 in) long. The plant rarely attains much height, preferring to grow prostrate and spread outwards.
Season Early summer.

Mimosa pudica

Sensitive plant; live and die; humble plant; shame plant

Family *Leguminosae* **Subfamily** *Mimosoideae*
Place of origin Tropical America
Description Widespread in both old and new world tropics, a small, short-lived perennial with spiny stems, from 30–90 cm (1–3 ft) long, which either grows prostrate or clambers over small obstacles. **Leaves:** alternate, small, long-stemmed, pinnate, with 15–20 pairs of oblong leaflets. These are extremely sensitive, the leaflets closing together and the whole leaf drooping when touched. **Flowers:** in round, rosy-mauve heads, about 12 mm (½ in) across.
Season Summer.
Remarks Often grown in cool climates as a curiosity, when it is treated as an annual and kept under glass.

Mirabilis jalapa (*M. uniflora*)

Four o'clock plant; marvel of Peru; beauty of the night

Family *Nyctaginaceae* **Place of origin** Tropical America
Description Herbaceous perennial often grown as an annual but when left produces heavy tuberous roots after the fashion of a dahlia. Height 60 cm (2 ft). **Leaves:** opposite, entire, ovate-lanceolate, about 8 cm (3 in) long, smooth. **Flowers:** opening late afternoon, hence common name, fragrant, flat-faced with 5 lobes and a 2·5–5 cm (1–2 in) tube behind. Colours variable, white, yellow, red or pink, often striped or mottled with other shades.
Season Summer, intermittently at other times.

Lotus berthelottii

Mimosa pudica

Mirabilis jalapa

Musa × paradisiaca

(*M. acuminata* × *M. balbisiana; M. × sapientum*)

Banana; plantain; plátano

Family *Musaceae* **Place of origin** Asia
Description Giant tree-like plants, clump-forming, with stout stems up to 6 m (20 ft) high. **Leaves:** arranged round the stems spirally, each up to 2·4 m (8 ft) long and 60 cm (2 ft) wide, broadly oblong, upright-growing but fraying in the wind, light green, sometimes purple-blotched. **Flowers:** on stout stems, red and purple with some yellow inside, upright, male and female blooms separate. As the fruits develop the increase in weight causes the stem to droop downwards. Bananas are always picked green and develop their colour in storage. The main plant dies after fruiting but is replaced by basal offsets. The ornamental species *Musa coccinea* (red banana) grows to 1·2 m (4 ft).
Season Any time of the year.

Nicolaia elatior (*Phaeomeria magnifica; P. speciosa; Alpinia magnifica; Amomum magnificum*)

Torch ginger; Philippine waxflower

Family *Zingiberaceae* **Place of origin** South east Asia, New Guinea
Description Perennials with striking flowers commonly cultivated in tropical countries, especially those with reasonable humidity. **Stems:** leafy, growing 3·6–4·5 m (12–15 ft). **Leaves:** oblong-lanceolate, resembling those of the banana but smaller, 60 cm (2 ft) long by about 15 cm (6 in) wide, smooth. **Flowers:** torch-like, singly on 1·2 m (4 ft) stems, brightly coloured, red waxy bracts with pink margins, red inner petals, 13 cm (5 in) across.
Season Spring.
Remarks Young shoots sometimes used in curries, aromatic roots for flavouring various dishes.

Osteospermum ecklonis (*Dimorphotheca ecklonis*)

Blue and white daisy bush
Family *Compositae* **Place of origin** South Africa
Description A shrubby, evergreen, bushy perennial, sometimes sprawling, to 60 cm (2 ft) or more. Becomes woody with age. **Leaves:** alternate, simple, oblong-lanceolate with sparsely toothed edges, around 5 cm (2 in) long. **Flowers:** glistening white daisies with deep blue centres, about 5–8 cm (2–3 in) across, reverse of flowers streaked blue and mauve.
Season Summer, but intermittently most of the year in warm climates.
Remarks Only opens freely in sunshine or good light.

Musa × paradisiaca

Musa coccinea

Nicolaia elatior

Osteospermum ecklonis

Oxalis pes-caprae (*O. cernua*) Ψ ※ ✳

Bermuda buttercup; buttercup oxalis; cape sorrel; English weed (Malta)

Family *Oxalidaceae* **Place of origin** South Africa
Description Low-growing herbaceous perennial forming scaly bulbils near soil surface. **Leaves:** basal, three-parted like clover on 13 cm (5 in) stems. **Flowers:** plentiful, making sheets of blossom, bright yellow, five-petalled, about 2–3 cm (1 in) across in loose umbels. Naked stems. A double form occurs, particularly in Madeira.
Season Spring, summer, intermittently at other times.

Pentas bussei (*P. coccinea*) Ψ ⊘ ◇

Star cluster

Family *Rubiaceae* **Place of origin** East tropical Africa
Description Low-growing shrubby plant, growing to 90 cm (3 ft).
Leaves: opposite, entire, ovate or oval-lanceolate, hairy beneath, stalked. **Flowers:** about 2·5 cm (1 in) across in terminal heads, after the style of herbaceous phlox but smaller, bright scarlet, five-lobed.
Season Summer.
Remarks *P. lanceolata* is similar but has pale purple flowers, although there are white, magenta-pink and lilac forms.

Pitcairnia staminea Ψ ⊘ ◇

Family *Bromeliaceae* **Place of origin** Brazil
Description A stemless plant with dense rosettes of foliage and racemes of irregular, showy flowers. **Leaves:** long, strap-shaped, 30–60 cm (1–2 ft) long, 6–12 mm ($\frac{1}{4}$–$\frac{1}{2}$ in) wide, bright green, smooth, on short stems. **Flowers:** arranged loosely on 30–45 cm (1–1$\frac{1}{2}$ ft) stems, bright red, about 5 cm (2 in) long.
Season Winter or early spring.

Saccharum officinarum Ψ ⊘ ✳

Sugar cane

Family *Gramineae* **Place of origin** East Asia
Description A giant grass cultivated over most of the tropics. Usually unmistakable forming large stands of grassy shoots up to 4·5 m (15 ft) tall. Stems leafy and renewed annually from underground rhizomes. **Leaves:** overlapping, long, narrow, about 60 cm (2 ft) long and 6 cm (2$\frac{1}{2}$ in) wide, grassy. **Flowers:** in dense, woolly plumes, about 60 cm (2 ft) long at tops of the stems.
Season Summer.
Remarks Sugar is obtained from the central stem tissues.

Oxalis pes-caprae

Pentas bussei

Pitcairnia staminea

Saccharum officinarum

Strelitzia reginae

Bird of paradise flower; crane flower; bird's tongue flower

Family *Strelitziaceae*   **Place of origin** South Africa
Description Well-known flowers often available in florist shops. Large, tender to frost, perennial herb with a woody trunk or rhizome. **Leaves:** basal, oblong-lanceolate, about 45 cm (1½ ft) long, and 15 cm (6 in) wide on smooth stems of about the same length. **Flowers:** grouped in a rigid, boat-shaped, terminal inflorescence with protruding petal-like segments resembling ships' sails. Large, 20 cm (8 in) purple and orange with a green base, the whole structure calling to mind a bird's head.
Season Spring.
Remarks The plant was named for Charlotte, wife of George III, who was a princess of the house of Mecklenburgh Strelitz.

Thunbergia erecta (*Meyenia erecta*)

King's mantle; bush clock vine

Family *Acanthaceae*   **Place of origin** Tropical Africa
Description Often seen as a low-growing plant, frequently creeping over the ground. Under good conditions, however, it grows to a 1·8 m (6 ft) shrub. **Leaves:** opposite, simple, ovate, about 8 cm (3 in) long with toothed edges. **Flowers:** solitary, tubular, dark blue with 5 lobes and an orange throat, the tube white or yellowish; length of blooms about 6 cm (2½ in). There is a white form called 'Alba'.
Season Summer.

Tillandsia usneoides

Spanish moss; grey beard

Family *Bromeliaceae*   **Place of origin** From southern United States (Florida) to Argentine
Description Interesting plant, very common in many areas where it festoons trees and even telegraph poles with long forked trails of grey-green mossy stems. These may be up to 6 m (20 ft) in length. **Stems:** slender with very narrow, needle-like, 2·5–5 cm (1–2 in) leaves. **Flowers:** insignificant, greenish, difficult to see.
Season All seasons.
Remarks Spanish moss was at one time in great demand for packing purposes or stuffing pillows, etc. It is still used in some places.

Strelitzia reginae

Tillandsia usneoides

Thunbergia erecta

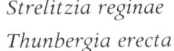

Tribulus cistoides
Puncture vine; calthrops

Family *Zygophyllaceae*   **Place of origin** Native of the Old World, now a pantropical weed.
Description Prostrate or creeping plant (rarely ascending), perennial (or sometimes annual) with tap root. **Leaves:** opposite, pinnate with 5 to 8 pairs of sessile leaflets, each about 12 mm ($\frac{1}{2}$ in) long but one in each pair usually larger than the other. **Flowers:** solitary, axillary, bright yellow, five-petalled flowers up to 2·5 cm (1 in) across. **Fruit:** round, separating when ripe into 4, (sometimes 6) sharp-barbed, spiny nutlets. These lodge in the feet of animals or fasten to skin or clothing. *T. terrestris* is similar but an annual with 2 bony spines in the upper part of the fruit and 2 small ones in the lower.
Season Most of the year.

Zantedeschia aethiopica
(*Z. africana; Richardia aethiopica*)
Arum lily; calla lily; pig lily

Family *Araceae*   **Place of origin** South Africa
Description Well-known plants with tuberous roots and spectacular flowers. Native to South Africa but now widely naturalized elsewhere, particularly in Western Australia. Height: 60–75 cm (2–2$\frac{1}{2}$ ft). **Leaves:** entire, arrow-shaped, basal, smooth, glossy, deep green, 25–45 cm (10–18 in) long, 10–20 cm (4–8 in) wide on long, smooth stems. **Flowers:** white arums with golden, poker-like spadix inside, 10–25 cm (4–10 in) long, slightly fragrant.
Season Spring, summer.

Zephyranthes candida
Zephyr lily; rain lily

Family *Amaryllidaceae*   **Place of origin** South America
Description Charming small, bulbous plants used in flower borders or for edging purposes in the tropics and subtropics. In cool climates they are sometimes grown in pots in greenhouses. **Leaves:** evergreen, basal, grassy, linear, somewhat fleshy, 20–30 cm (8–12 in) long. **Flowers:** profuse, singly on slender, 10–20 cm (4–8 in) stems, crocus-like, funnel-shaped, 6 segments, white or sometimes tipped with rose, about 5 cm (2 in) long. There are cultivars with larger flowers like 'Major'; *Z. grandiflora* has pink blooms, 10 cm (4 in) across; and *Z. tubiflora* the fire lily, deep orange flowers up to 5 cm (2 in) across.
Season Autumn.

Tribulus cistoides

Zephyranthes candida

Zantedeschia aethiopica

Index

Page numbers in **bold type** *denote colour photographs.*

135

Crete - 1989
Bouganvillea
Hibiscus
Agave
Plumbago
Opuntia

Morning Glory
Eucalyptus
Ficus
Oleander

Rose periwinkle
Mimosa
Eucalyptus
Ficus

On LANZAROTE - 1986 -

Acacia, Araucaria (Norfolk Island Pine),
Aloe (torch plant), Callistemon (Bottle brush), Cassia alata (Christmas candle),
Euphorbia milii (Crown of thorns),
Hibiscus rosa-sinensis (China rose),
Opuntia vulgaris (Prickly pear),
Yucca (Spanish bayonet),
Bougainvillea (Paper flower)
Solanum jasminoides (Chalice vine)
Agave (American aloe)
Geranium maderense (Madeira Stork's Bill)
Ipomoea acuminata (Morning glory)
Aeonium arboreum
Thunbergia erecta (King's mantle)
Acalypha wilkesiana (Jacob's coat)
Plumbago auriculata (Plumbago)
Agave attenuata
    americana (maguey)
Catharanthus roseus (rose periwinkle)
Eucalyptus (gum tree)
Ficus (fig tree)
Oleander
Dracaena draco (dragon tree)

Murcia - 1990

Acacia (mimosa), Aloe
Euphorbia (crown of thorns)
Hibiscus (china rose)
Opuntia vulgaris (prickly pear)
Yucca (Spanish bayonet
Bougainvillea (paper flower)
Agave (American Aloe)
Ipomoea (morning glory)
Plumbago
Eucalyptus
Ficus
Oleander
Pomegranate, lemon, orange, lime.
~~Aura~~ Araucaria (Norfolk island pine)
~~Coconut~~ palm (date)
Parkinsonia (Jerusalem thorn)
Canna indica
Osteospermum echlonis (blue & white
            daisy bush)
Bignonia (Clytostoma callistegioides)
        (Argentine trumpet vine)
Lantana camara (Shrub verbena)
Clianthus puniceus (Parrot's bill)
Solanum jasminoides (Potato vine)
Hottentot Fig — Carpobrotus edulis